FRP DECK
AND
STEEL GIRDER
BRIDGE
SYSTEMS

Analysis and Design

Composite Materials: Design and Analysis

Series Editor

Ever J. Barbero

FRP DECK AND STEEL GIRDER BRIDGE SYSTEMS

Analysis and Design

Julio F. Davalos
An Chen
Pizhong Qiao

CRC Press
Taylor & Francis Group
Boca Raton London New York

CRC Press is an imprint of the
Taylor & Francis Group, an **informa** business

CRC Press
Taylor & Francis Group
6000 Broken Sound Parkway NW, Suite 300
Boca Raton, FL 33487-2742

First issued in paperback 2017

© 2013 by Taylor & Francis Group, LLC
CRC Press is an imprint of Taylor & Francis Group, an Informa business

No claim to original U.S. Government works
Version Date: 20121220

ISBN 13: 978-1-138-07718-8 (pbk)
ISBN 13: 978-1-4398-7761-6 (hbk)

Library of Congress Cataloging-in-Publication Data

Davalos, Julio F.
   FRP deck and steel girder bridge systems : analysis and design / authors, Julio F. Davalos, An Chen, Pizhong Qiao.
      pages cm. -- (Composite materials: design and analysis)
      Includes bibliographical references and index.
      ISBN 978-1-4398-7761-6 (hardback)
      1. Bridges--Floors--Materials. 2. Fiber-reinforced concrete. 3. Polymer-impregnated concrete. I. Title.

TG325.6.D38 2013
624.2'83--dc23   2012049380

Visit the Taylor & Francis Web site at
http://www.taylorandfrancis.com

and the CRC Press Web site at
http://www.crcpress.com

Contents

Series Preface

Fifty years after their commercial introduction, composite materials have widespread use in many industries. Applications such as aerospace, windmill blades, highway bridge retrofit, and many more require designs that assure safe and reliable operation for 25 years or more. Using composite materials, virtually any property, including stiffness, strength, thermal conductivity, and fire resistance, can be tailored to the user's needs by selecting the constituent material, its proportion and geometrical arrangement, and so on. In other words, the engineer is able to design the material concurrently with the structure. Also, modes of failure are much more complex in composites than in classical materials. Such demands for performance, safety, and reliability require that engineers consider a variety of phenomena during the design. Therefore, the aim of the Composite Materials: Design and Analysis Book Series is to bring to the design engineer a collection of works written by experts on every aspect of composite materials that is relevant to the design.

Variety and sophistication of material systems and processing techniques have grown exponentially in response to ever-increasing numbers and types of applications. Given the variety of composite materials available, as well as their continuous change and improvement, understanding of composite materials is by no means complete. Therefore, this book series serves not only the practicing engineer but also the researcher and student who are looking to advance the state of the art in understanding material and structural response and develop new engineering tools for modeling and predicting such responses.

Thus, the series is focused on bringing to the public existing and developing knowledge about the material–property relationships, processing–property relationships, and structural response of composite materials and structures. The series scope includes analytical, experimental, and numerical methods that have a clear impact on the design of composite structures.

Ever J. Barbero
West Virginia University, Morgantown

Preface

In recent years, the demands in civil infrastructure have provided opportunities for the development and implementation of fiber-reinforced polymer (FRP) decks, in both rehabilitation projects and new constructions, due to their reduced weight and maintenance costs as well as their enhanced durability and service life. Numerous bridges with FRP decks have been built all over the world. The number is still increasing due to the confidence gained from past examples, and new areas, such as applications in cold regions, are still being identified. Unlike other types of conventional decks, FRP bridge panels are usually more complicated in the form of sandwich or cellular shapes, which makes it challenging to predict their behavior. Therefore, significant research, both theoretical and experimental, has been carried out since the 1990s, including the work by the authors and other collaborators.

In contrast to the increasing applications and extensive research on FRP bridge decks, the design process is still at a preliminary stage and not in a code format. Individual decks and bridges are designed on a job-by-job basis, usually using a combination of sample testing, numerical finite element technique, and empirical judgment. Apparently, such approaches have hampered the wider acceptance of relatively new and innovative FRP bridge decks. Therefore, in order to promote wider acceptance and sound technical bases for FRP decks, there is a need to develop standards and guidelines, similar to those for other types of decks as detailed in the American Association of State Highway and Transportation Officials (AASHTO) specifications, which can be readily used by design engineers.

The aim of this book is to address these concerns. It presents the analysis and design of FRP bridge decks, illustrated specifically for a honeycomb FRP (HFRP) sandwich deck and steel girder bridge system. Three complementary parts are included: FRP deck, shear connectors between the deck and steel girders, and behavior of the bridge system. The section on the FRP deck focuses on (1) stiffness evaluation considering sinusoidal core geometry, general core configuration, and skin effect (Chapter 2), and (2) strength evaluation for out-of-plane compression, out-of-plane shear, and facesheet laminates (Chapter 3). The section on shear connector focuses on its strength, stiffness, and fatigue performance (Chapter 4). At the bridge system level, studies are provided for effective flange width and load distribution factor for a FRP deck and steel girder bridge system (Chapter 5). Design guidelines are proposed based on the findings from Chapters 2 through 5, and design examples are provided to illustrate the applications of these design guidelines (Chapter 6). For completeness, a systematic analysis and design approach for single-span FRP deck-stringer bridges with a design example is also provided (Chapter 7). While the guidelines for FRP decks are focused on a specific sandwich panel, they can be adapted to other types of decks, and the guidelines for a shear connector and bridge system can be used for other types of FRP decks.

The unique feature of this book is a combination of analytical models, numerical analyses, and experimental investigations to formulate a practical analysis procedure, based upon which design formulations are proposed. Many new or improved theories are presented for the first time, including

- Homogenization theory to predict the stiffness of sandwich panels with sinusoidal core and furthermore with general configurations
- Skin effect, which can be decomposed as bending and shear warping effects, on the stiffness and strength of sandwich panels
- Buckling behavior of core sheets with elastically restrained loaded edges
- Experimental method to quantify the coefficient of elastic restraint between the core and facesheet of sandwich panels
- Behavior of bridge system with partial degree of composite action

Also, for the first time a complete set of design guidelines is proposed, including an FRP deck, shear connector, and bridge system. The unique combination of analytical and practical features makes this book useful for both professionals in the bridge industry, including design engineers and Department of Transportation officials who are interested in the application of FRP bridge decks, and researchers who are interested in the behavior of sandwich structures.

Acknowledgments

The contributions of the following colleagues, friends, and former students are gratefully acknowledged: Dr. Ever J. Barbero, Dr. Karl E. Barth, Dr. Jennifer R. McConnell, Dr. Roberto Lopez Anido, Dr. Indrajit Ray, Dr. Hani A. Salim, Dr. X. Frank Xu, and Dr. Bin Zou; and numerous graduate students at West Virginia University, the University of Akron, the University of Idaho, and Washington State University.

In particular, this book is dedicated to Dr. Jerry D. Plunkett, president and CEO of Kansas Structural Composites, Inc. and Missouri Structural Composites LLC. His creative inventiveness of the honeycomb FRP deck, his invaluable technical contributions to the field, and his generous willingness to work collaboratively with us has made it possible for this work to be accomplished, and for this book to be published. Dr. Plunkett has been a continued source of inspiration for our joint work, which includes raceway systems for fish production; novel panel products for building walls, floors, and roofs; FRP enclosures for bridge piers and abutments; and blast and ballistic protection of infrastructures. We sincerely appreciate his creativity, wisdom, and sincere friendship.

We are grateful to our loving families for their constant support and affection.

> To my son, Julio Gustave, his wife, Leora, and my grandson, Julio Joaquin.
>
> **Dr. Davalos**

> To my parents, Yuntong and Shulan, my wife, Lili Yang, and my daughters, Angelina and Laura.
>
> **Dr. Chen**

> To my parents, Hengkuan and Xiufang, my wife, Dan Liu, and my daughters, Lily Danni and Lynnlin Jennifer.
>
> **Dr. Qiao**

The following agencies provided financial support to this study: National Science Foundation–Partnerships for Innovation program (NSF-PFI: EHR-0090472), NCHRP-IDEA program, West Virginia Department of Transportation—Division of Highways, West Virginia Higher Education–Research Challenge Grant, and the University of Akron.

In loving memory of Cristina Martha Lozada-Davalos and Virginia Ruth Lee Wilson-Plunkett.

About the Authors

Dr. Julio F. Davalos initially obtained three undergraduate degrees in civil engineering, engineering mechanics, and construction management, and later BS, MS, and PhD degrees from Virginia Tech, where he was recognized as Outstanding Young Alumnus in Civil Engineering in 1989, and Outstanding Alumnus in April 2011. His expertise is in mechanics and structural engineering, and his research work includes theoretical and experimental studies on advanced materials and systems. His work is directed to civil infrastructure rehabilitation and protection, with particular emphasis on highway bridges, buildings, and mass transit tunnels. He was the Benedum Distinguished Teaching Professor and center director at West Virginia University, and he is currently professor and chair of the Department of Civil Engineering at the City College of New York–CUNY. He is a devoted teacher engaged in the development of innovative teaching methods and technologies. Dr. Davalos has been honored with over 60 WVU/state/national awards for teaching, research, and innovative designs and concepts, and he has several patent applications in materials and structures.

His publications record of about 300 articles includes a 1996 Best Design Paper Award from the Composites Institute's 51st Annual Conference, and the 1997 Best Research Paper Award from the *American Society of Civil Engineers Journal of Composites for Construction*. In 1993, Dr. Davalos was invited by the U.S. Senate Subcommittee on Science Technology and Space to provide expert testimony on advanced composites in civil infrastructure. Dr. Davalos's coauthored paper, "Step-by-Step Engineering Design Equations for FRP Structural Beams," was awarded the Best Design Paper Award and Overall Best Paper Award (*Modern Plastics Magazine*) from the Composites Institute of SPI, May 1999. He was also honored with the Applied Research Paper Award from the ASCE Construction Institute (2002).

Dr. Davalos has produced bridge design manuals for state and federal agencies, has taught short courses on composite materials and the design of timber bridges to practicing engineers, and has developed design equations and guidelines for FRP structural shapes, composite decks, and bonded FRP reinforcements. The WVU College of Engineering and Mineral Resources selected Dr. Davalos as outstanding researcher ten times, and researcher of the year five times. He was also named numerous times as outstanding teacher and teacher of the year. In 1998, he was recognized as professor of the year in West Virginia—the highest honor for a faculty member in the state.

In the last 15 years, and after being honored with the WVU Foundation Outstanding Teaching Award in 1995 as assistant professor, Dr. Davalos has demonstrated continuous growth and enthusiasm in research and teaching innovations. He has developed and implemented unique and effective active learning teaching methods and has energized students and faculty to reach new levels of learning and teaching excellence. In 2005, he was honored for a second time with the WVU Foundation Outstanding Teaching Award. Dr. Davalos was honored with the WVU–CE Department's Excellence in Teaching Award every year from 1994 to 2011.

Dr. An Chen, assistant professor of civil engineering at the University of Idaho, received his BE and MS degrees in civil engineering from Dalian University of Technology, China, and his PhD from West Virginia University. Prior to joining the University of Idaho, he was a research assistant professor at West Virginia University and junior associate at the Office of James Ruderman, LLP. He is a licensed professional engineer in civil engineering and Leadership in Energy and Environmental Design® Accredited Professional (LEED® AP).

Dr. Chen's research background is in sustainable structural engineering, covering advanced materials, interface bond and fracture mechanics, and applied mechanics. His research can be broadly categorized into two areas: (1) green buildings—focused on developing energy-efficient structures and sustainable materials using recycled materials, and (2) sustainable civil infrastructure—with emphasis on fast-construction bridges and developing sustainable rehabilitation systems for deteriorated bridges. He has worked on many research projects funded by the university and state and federal agencies. He has three pending patents. Dr. Chen's publications record includes about 60 refereed journal and conference papers and project reports. He is a committee member of ASCE and a member of many other professional organizations, including AISC, ASC, and IIFC. He has been chair of symposia and technical sessions, and a member of scientific committees for several international conferences. Dr. Chen also has extensive industrial experience. As a project manager working in New York City, he has completed designs of numerous new and renovation projects for high-rise and middle-rise buildings.

Dr. Pizhong Qiao, professor of civil and environmental engineering at Washington State University (WSU), and founder of Integrated Smart Structures, Inc. (Copley, Ohio), received his PhD in civil engineering (advanced materials, solid mechanics, and structures) from West Virginia University (WVU) in 1997. Before joining WSU, Dr. Qiao was an assistant/ associate professor of civil engineering at the University of Akron, Ohio, from 1999 to 2006 and a research assistant professor of civil and environmental engineering at WVU from 1997 to 1999. He is a registered professional engineer (PE) in structural engineering and certified in the practice of structural engineering from the Structural Engineering Certification Board (SECB). He was named a Fellow of the American Society of Civil Engineers (ASCE) (April 2007) in recognition of "his significant contributions to civil engineering research, application, and teaching that have advanced the state-of-the-art in theory and practice in the composites area."

Dr. Qiao has worked extensively in development, research, and application of advanced and high-performance materials (smart materials, polymer composites, and sustainable concrete) in civil and aerospace engineering. His research interests include analytical and applied mechanics, smart and composite materials, interface mechanics and fracture, impact mechanics and high-energy absorption materials, structural health monitoring, integrated intelligent structural systems, materials characterization, and sustainable concrete. His extensive publication record includes about 300 technical publications (several book chapters, 132 international journal articles, and more than 160 conference proceedings papers/presentations). He is one of the most highly cited scientists (about the top 1%) in the field of engineering according to Essential Science Indicators (ESI). He also holds one patent on bistable bond lattice structures for blast-resistant armor appliqués. Dr. Qiao won the Best Paper Award twice from the Composites Institute and was the recipient of the Best Technical Paper Award from *Modern Plastics Magazine* and the inaugural Best Research Paper Award from the *ASCE Journal of Composites for Construction*. He is the 2005 Outstanding Researcher of the Year and the 2006 Outstanding Teacher of the Year in the College of Engineering at the University of Akron. He was selected as the 2007 Changjiang Distinguished Scholar by the Ministry of Education of the People's Republic of China and honored with the 2007 Outstanding Technical Contribution Award from the ASCE Aerospace Division for his contribution in the area of fracture mechanics of layered structures. Most recently, Dr. Qiao was honored with the 2012 Anjan Bose Outstanding Researcher Award from the College of Engineering and Architecture, WSU.

Dr. Qiao is the executive committee member of Aerospace Division, ASCE (2010–2015). He serves as editor-in-chief of *Frontiers in Aerospace Engineering*, associate editor of three major journals (*International Journal of Structural Health Monitoring*, *ASCE Journal of Engineering Mechanics*, and *ASCE Journal of Aerospace Engineering*), editorial board member of eight journals of international circulation (*Journal of Advanced Materials*, *Disaster Advances*, etc.), and guest editor (2005, 2008, and 2011) of the *ASCE Journal of Aerospace Engineering*. He has served as a reviewer for more than 60 journals of international circulation, and has been chair for numerous sessions/symposia in international conferences. Dr. Qiao is chair of Aerospace Division Advanced Materials and Structures Committee (2005 to present) and past chair of ASCE Engineering Mechanics Institute Stability Committee (2009–2011). He is also a member of the ASCE Structural Engineering Institute (SEI) Technical Committee on Timber Bridges.

1

Introduction

1.1 Background

According to a report from Market Development Alliance of the FRP Composites Industry (MDA 2003), today's bridge owners are faced with unique challenges as a result of a severely deteriorating infrastructure, insufficient funding, and a demanding public. A released study (Ellis 2011) funded by the Federal Highway Administration (FHWA) estimates the annual direct cost of corrosion for highway bridges to be $6.43 billion to $10.15 billion. This includes $3.79 billion to replace structurally deficient bridges over the next 10 years and $1.07 billion to $2.93 billion for maintenance and cost of capital for concrete bridge decks. In addition to these direct costs, the study's life cycle analysis estimates indirect costs to the user due to traffic delays and lost productivity at more than 10 times the direct cost of corrosion. Although most bridge owners continue to make decisions based on lower initial cost, it has become apparent that this approach does not work, and in the near future more money will be spent maintaining existing structures than building new ones. As a result, there are significant opportunities for fiber-reinforced polymer (FRP) bridge decks that are corrosion resistant, and can be rapidly installed.

Numerous bridges with FRP decks have been built in the world. In the United States, federal technology transfer initiatives taken to utilize composite manufacturing capacities by the military and aerospace industries have led to the proliferation of FRP use in the bridge industry. Some companies capitalized on the potential of the transportation market and helped advance the use of FRP on bridge structures. Based on NCHRP Report 564 (2006), more than 100 bridges with FRP bridge decks have been built since the mid-1990s, and they are currently in good condition in general. The number is still increasing with the maturing of the technology.

Primary benefits of FRP decks include durability, light weight, high strength, rapid installation, lower or competitive life cycle cost, and high-quality manufacturing processes under controlled environments. Compared with cast-in-place concrete decks, FRP bridge decks typically weigh 80% less, can be erected twice as fast, and have service lives that can be two to three times greater. Although, based on initial in-place material cost, FRP bridge decks typically cost two to three times more than a conventional deck, life cycle costs, light weight, and rapid installation tend to be features that justify the use of FRP bridge decks.

FRP bridge decks commercially available at the present time can be classified according to two types of construction: sandwich and adhesively bonded pultruded shapes (Bakis et al. 2002). For sandwich constructions, cellular geometries are the most efficient use of core materials for weight-sensitive applications. Due to the ease with which facesheets and core materials can be changed in manufacturing, sandwich construction presents great flexibility in designing for varied depths and deflection requirements. Facesheets of sandwich bridge decks are primarily composed of E-glass mat or roving infused with polyester or vinylester resins. Current core materials are rigid foams of thin-walled cellular FRP materials. Figure 1.1 displays a honeycomb fiber-reinforced polymer (HFRP) panel, with sinusoidal core configuration in the plane extending vertically between face laminates, which was introduced for highway bridge decks by Plunkett (1997). More recently, a sandwich panel with I-beam-type core configuration (ZellComp® system; Figure 1.2) was developed by ZellComp, Inc. This system was installed on the Morrison Bridge in Portland, Oregon, and is the largest FRP bridge deck in the United States to date.

Adhesively bonded pultruded shapes can be economically produced in continuous lengths by numerous manufacturers using well-established processing methods. Design flexibility on this type of deck is obtained by changing the constituents of the defined shapes (such as fibers and fiber

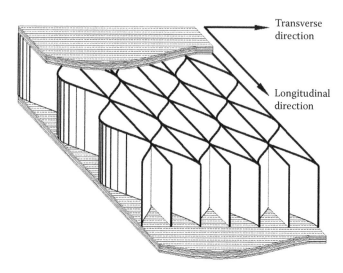

Transverse direction

Longitudinal direction

FIGURE 1.1
HFRP panels with sinusoidal core configuration (KSCI).

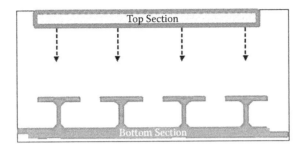

FIGURE 1.2
ZellComp® decking system. (From http://www.zellcomp.com/technology_n_markets.html. With permission.)

orientations), but changes in geometry are very difficult to implement. Several decks constructed with pultruded shapes are shown in Figure 1.3. The pultruded shapes are typically aligned transverse to the traffic direction over supporting stringers. Each deck design has advantages in terms of stiffness, strength, and field implementation. In laboratory testing, the observed failures in such decks are generally by local punching shear and crushing or large-scale delamination of the shapes constituting the cross section.

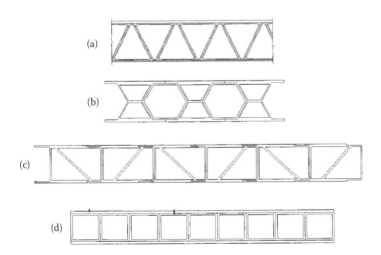

FIGURE 1.3
FRP decks produced from adhesively bonded pultruded shapes: (a) EZSpan (Atlantic Research), (b) Superdeck (Creative Pultrusions), (c) DuraSpan (Martin Marietta Materials), and (d) square tube and plate deck (Strongwell). (From Bakis, C. E. et al., *ASCE Journal of Composites for Construction*, 6(2), 73–87, 2002.)

1.2 Implementation of HFRP Sandwich Deck Panels in Highway Bridges

A technical comparison of sandwich and pultruded decks is shown in Table 1.1, from which we can see that the honeycomb FRP (HFRP) sandwich panels provided by Kansas Structural Composites, Inc. (KSCI) excel in terms of weight, cost, and deflection among all commercial FRP decks. In addition, the flexibility of the manufacturing process permits custom production of panels of any depth, while a pultruded section has a fixed geometry dictated by the forming steel die used. The HFRP sandwich deck panels have been extensively used in both rehabilitation projects and new constructions, with several examples shown below.

Initially, the HFRP sandwich panel (Figure 1.1) was developed by KSCI for a short-span bridge deck superstructure, without supporting stringers, located near Russell, Kansas. It is approximately 7 m (23 ft) long by 8.53 m (28 ft) wide and consists of three 55.88 cm (22 in.) thick longitudinal panels, with the sinusoidal core waves oriented along the traffic direction. The two exterior panels had the standard W-beam guardrail preinstalled (see Figure 1.4), and the wearing surface consisted of a 9.5 mm (3/8 in.) coarse aggregate polymer concrete that was co-cured with the top face of the panel at the manufacturing plant. The bridge superstructure was installed in 8 h. This bridge was designed for an American Association of State Highway and Transportation Officials (AASHTO) HS-20 truck loading and successfully load tested by the Kansas Department of Transportation. This bridge is the first all-FRP bridge installed on a public road in the United States.

In 1999, the concrete decks of two identical 13.72 m (45 ft) span steel stringer bridges in Crawford County, Kansas, were replaced with the KSCI honeycomb panels. Because of the close 0.69 m (27 in.) spacing of the W21x68 steel

TABLE 1.1

Summary of Deck Characteristic for Two Fabrication Methods

Deck System	Depth (in.)	lb/ft^2	Dollar/ft^2	Deflection (reported)	Deflection (normalized)
Sandwich construction					
Hardcore composites	7–28	20–23	53–110	L/785	L/1,120
KSCI	5–24	15	65	L/1,300	L/1,300
Adhesively bonded pultrusions					
DuraSpan	7.5	18	65-75	L/450	L/340
SuperDeck	8	22	75	L/530	L/530
EZSpan	9	20	80–100	L/950	L/950
Strongwell	5–8	23	65	L/605	L/325

Source: From Bakis, C.E. et al., *ASCE Journal of Composites for Construction*, 6(2), 73–87, 2002.

FIGURE 1.4
FRP sandwich deck with attached guardrail.

stringers, the transverse FRP panels were only 0.12 m (4.75 in.) thick, and to achieve the required elevation of the replaced concrete deck, honeycomb FRP "saddle" beams were placed as spacers over the steel stringers by means of an inverted FRP channel, which served as the bottom facesheet of the sandwich assembly (see Figure 1.5). These spacers allow the use of more economical thinner deck panels to accomplish the required grade and profile. These bridges were designed for an AASHTO HS-25 truck loading.

The Hanover Bridge deck replacement project was completed in 2001 in West Virginia. The project consisted of two simply supported 17.07 m (56 ft) spans. The existing 8.53 m (28 ft) wide nail-laminated timber deck over steel stringers was replaced with an 0.2 m (8 in.) thick FRP honeycomb deck, which was installed over one-half of the bridge at a time to avoid traffic interruption during construction. A 6.35 mm (1/4 in.) neoprene bearing pad was placed over each stringer, and a 3.18 mm (1/8 in.) thick inverted FRP channel

FIGURE 1.5
Placement of FRP sandwich deck.

was secured over the bearing pad. The 4.27 m (14 ft) long half panels were placed transversely over the stringers and joined at 2.44 m (8 ft) intervals using a FRP box beam insert, which also served to connect the panels to the supporting stringers by a steel clamp-bar mechanism that secures the deck to the top flange of the stringer. In this project, the preinstalled polymer concrete surface was covered with a hot-mix asphalt to create required crown and superelevation.

Recently, extensive studies have been conducted on the HFRP deck panels subjected to low temperature (Nordin et al. 2010; Qiao et al. 2011). Further implementations in cold regions, such as Alaska and Washington state, are expected.

1.3 Objectives

Thus far, the design process for sandwich decks is not in a code format. Rather, individual decks are designed on a job-by-job basis since the bridge decks have been built using proprietary systems as shown in Figures 1.1 to 1.3. The development of standards and guidelines is needed in order to promote wider acceptance of composite sandwich panels in construction. Characterizations of stiffness and strength properties of FRP decks, and FRP deck-girder bridge systems, are necessary to facilitate the development of design guidelines, which is the objective of this book. In particular, this book is focused on stiffness/strength evaluations for FRP deck panels and FRP deck-girder bridge systems, with a focus on HFPR sandwich panels with sinusoidal core geometry (Figure 1.1) because of its favorable advantages, as described in Section 1.2. The information presented includes both experimental investigations and theoretical analyses. While the studies for FRP decks are focused on a specific sandwich panel, they can be adapted to other sandwich configurations, and the studies for shear connections and bridge systems can be used for other types of FRP decks.

1.4 Organization

There are seven chapters in this book. Chapter 1 includes the introduction and organization of the book. Chapter 2 is focused on stiffness evaluation of the FRP sandwich decks with sinusoidal core geometry, with general core configuration, and with skin effect. Equivalent orthotropic properties are developed representative of the complex honeycomb geometry. Equivalent properties of face laminates are obtained using micromechanics models, and the effective orthotropic properties of honeycomb core are obtained from a homogenization process using a combined energy method and mechanics of materials approach. A simplified analysis procedure is presented that can be used in design applications.

Strength evaluation of the FRP deck with sinusoidal core geometry is presented in Chapter 3. Core materials for sandwich structures are primarily subjected to out-of-plane compression and shear, and the facesheet laminates sustain mainly membrane forces due to bending. Therefore, strength evaluations are presented for out-of-plane compression, out-of-plane shear, and facesheet laminates, through a combination of analytical models, numerical analyses, and experimental investigations.

Chapter 4 is focused on a mechanical shear connector. A prototype mechanical shear connector is presented that can accommodate any panel height for any type of FRP deck. Static and fatigue tests were conducted on push-out connection specimens, which were subsequently evaluated experimentally on an actual scaled bridge model. The strength, stiffness, and fatigue performance characteristics of the connection are investigated.

Chapter 5 discusses FRP deck–steel girder bridge systems with special considerations for load distribution factors and effective flange width considering partial composite action. Extensive tests were conducted on a one-third-scaled bridge model. The scaled bridge test specimen consisted of an FRP sandwich deck attached to steel stringers by the mechanical connector described in Chapter 4. A harmonic analysis is formulated to define effective width for FRP decks as an explicit solution.

Design guidelines are presented in Chapter 6 based on findings from Chapters 2 to 5. Examples are given to illustrate the use of the recommended guidelines to design the FRP deck, shear connector, and bridge system.

Finally, Chapter 7 presents a systematic analysis and design approach for single-span FRP deck-stringer bridges with a design example.

References

Bakis, C.E., Bank, L.C., Brown, V.L., Cosenza, E., Davalos, J.F., Lesko, J.J., Machida, A., Rizkalla, S.H., and Triantafillou, T.C. (2002). Fiber-reinforced polymer composites for construction-state-of-art review. *ASCE Journal of Composites for Construction*, 6(2), 73–87.

Ellis, Z. (2011). *Corrosion cost and preventive strategies in the United States*. FHWA-RD-01-156. Retrieved from http://www.fhwa.dot.gov/publications/publicroads/02may/newpubs.cfm.

MDA. (2003). Retrieved from http://www.mdacomsoites.org/psbridge_vehicular_print.html.

NCHRP Project 10-72. (2009). Bridge deck design criteria and testing procedures. Retrieved from http://144.171.11.40/cmsfeed/TRBNetProjectDisplay.asp?Project ID=292.

NCHRP Report 564. (2006). Field inspection of inservice FRP bridge decks. Transportation Research Board, Washington, DC.

Nordin, C., Ma, Z., and Penumadu, D. (2010). Combined effect of loading and cold-temperature on the stiffness of glass fiber composites. *ASCE Journal of Composites for Construction*, 14(2), 224–230.

Plunkett, J.D. (1997). *Fiber-reinforced polymer honeycomb short span bridge for rapid instal-lation*. IDEA Project Report.

Qiao, P., Fan, W., Husley, J. L., and McLean, D. (2011). *Smart FRP composite sandwich bridge decks in cold regions*. Report 107018 for Alaska University of Transportation Center and Alaska Department of Transportation and Public Facilities.

ZellComp, Inc. (2011). Retrieved from http://www.zellcomp.com/technology_n_markets.html.

2

FRP Deck: Stiffness Evaluation

2.1 Stiffness of FRP Honeycomb Sandwich Panels with Sinusoidal Core

2.1.1 Introduction

This section presents an approximate analytical solution verified by finite element modeling and experimental testing of fiber-reinforced polymer (FRP) honeycomb panels with sinusoidal core (Davalos et al. 2001).

In the aerospace industry, a high degree of refinement in theoretical modeling and analysis of honeycomb structures has been achieved through its relatively long history of applications (Noor et al. 1996), with particular attention devoted to research and development of sandwich panels. The computational models available for sandwich panels and shells were reviewed by Noor et al. (1996), and numerous references were provided. The focus of this section is on modeling of a unique FRP honeycomb sandwich structure with sinusoidal core geometry; the goal is to develop equivalent elastic properties for the core structure in order to significantly simplify the modeling process of the sandwich panel. The early research in this area includes the work of Kelsey et al. (1958), where the out-of-plane shear stiffness of a hexagonal honeycomb with higher/lower bounds was evaluated by the classical energy method. Gibson and Ashby (1988) presented the predictions for both in-plane and out-of-plane shear stiffness properties of a hexagonal honeycomb using mechanics of materials and energy methods. Another relatively new technique is homogenization theory (Shi and Tong 1995a), which is generally effective and accurate for estimation of honeycomb properties.

For simplified analysis and design optimization of FRP sandwich structures, it is useful to define equivalent orthotropic properties representative of the complex honeycomb geometry. In this section, an analytical solution is presented for the evaluation of equivalent properties of a honeycomb core with sinusoidal wave configuration, which is based on energy method and mechanics of materials approach. The material properties of the face laminates and core walls are predicted by micro/macromechanics models (Davalos et al. 1996). With the equivalent properties proposed in this study, the responses of both sandwich beams and panels are evaluated analytically and verified numerically and experimentally. Also, a brief summary is given of the field implementation of the sinusoidal core sandwich panel for bridge deck applications.

2.1.2 Modeling of FRP Honeycomb Panels

In this section, using a micro/macromechanics approach for face laminates and an energy method combined with a mechanics of materials approach for honeycomb core, the modeling of equivalent properties for face and core elements is presented.

2.1.2.1 Geometry of Honeycomb Core

The FRP panel for practical civil engineering applications (such as bridge decks) considered in this study was developed by Kansas Structural Composites, Inc. (KSCI) (Plunkett 1997). The geometry of the sandwich structure is intended to improve stiffness and buckling response by the continuous support of core elements with the face laminates. Originated from the basic concept of sandwich structures, two faces composed of FRP laminates are co-cured with the core as shown in Figure 2.1. The core geometry consists of closed honeycomb-type FRP cells. It is noteworthy that the thermosetting property of resin distinguishes honeycomb cores from their metal counterparts in both manufacturing and consequent corrugated shapes. Unlike traditional metal sandwich structures, the shape of the FRP corrugated cell wall is defined by a sinusoidal function in the plane. The combined flat and waved FRP cells are produced by sequentially bonding a flat sheet to a corrugated sheet, which is similar to the processing of the section resin sandwich

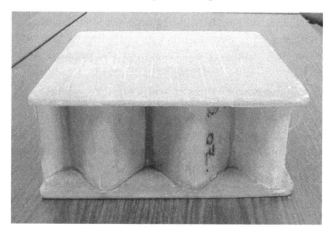

FIGURE 2.1
FRP honeycomb sandwich panel.

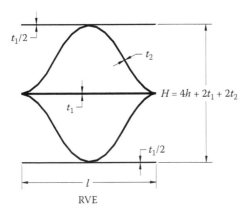

FIGURE 2.2
Representative volume element (RVE) of honeycomb core.

panel. The assembled cellular core is then co-cured with the upper and bottom face laminates to build a sandwich panel (see Figure 1.1).

The waved flutes or core elements are produced by forming FRP sheets to a corrugated mold. As shown in Figure 2.2, the distance of adjoining crests represents the wavelength l, and the interval between two adjoining flats gives the amplitude $2h$. In the coordinate system of Figure 2.2, the wavefunction of a corrugated core wall can be defined as

$$y = h(1 - \cos\frac{2\pi x}{l}) \tag{2.1}$$

and the dimensions of the sinusoidal core laminate are $h = 25.4$ mm (1 in.) and $l = 101.6$ mm (4 in.). The constituent materials used for the honeycomb sandwich panel (both face laminates and core) consist of E-glass fibers and polyester resin, and their properties are listed in Table 2.1.

2.1.2.2 Modeling of Face Laminates

The apparent engineering properties of laminated panels can be predicted by a combined micro- and macromechanics approach (Davalos et al. 1996). The prediction of ply properties by micromechanics is well defined (Chamis 1984), and the stiffness properties of each layer can be computed from

TABLE 2.1

Properties of the Constituent Materials

Material	E, GPa (×10⁶ psi)		G, GPa (×10⁶ psi)		v	ρ, g/cm³ (lb/in.³)
E-glass fiber	72.4	(10.5)	28.8	(4.18)	0.255	2.55 (0.092)
Polyester resin	5.06	(0.734)	1.63	(0.237)	0.30	1.14 (0.041)

existing micromechanics models such as rule of mixtures (ROM) (Jones 1999), periodic microstructure (PM) (Luciano and Barbero 1994), and composite cylinders (CCs) (Hashin and Rosen 1964), where each layer is modeled as a homogeneous, linearly elastic, and generally orthotropic material. Based on ply properties and lay-up, the apparent stiffness of the face laminate can be predicted by classical lamination theory.

2.1.2.2.1 Microstructure—Lay-Up

A typical face laminate may include the following four types of fiber layers (Figure 2.3):

1. Chopped strand mat (ChSM), which is made of short fibers randomly oriented, resulting in nearly isotropic in-plane properties. This layer is placed at the inner face of the laminate and provides a uniform and resilient bond between the face plate and the core.
2. Continuous strand mat (CSM), consisting of continuous randomly oriented fibers. This product is commonly used as backing material for nonwoven fabrics and can be modeled as an isotropic layer.
3. Bidirectional stitched fabrics (SFs) with balanced off-angle unidirectional fibers (e.g., 0°/90° or ±45°).
4. Unidirectional layer of fiber bundles or rovings.

In general, the fiber architecture of upper and bottom face laminates is symmetric about the mid-surface plane of the sandwich panel, while each

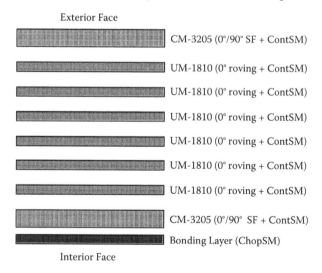

Exterior Face

CM-3205 (0°/90° SF + ContSM)

UM-1810 (0° roving + ContSM)

UM-1810 (0° roving + ContSM)

UM-1810 (0° roving + ContSM)

UM-1810 (0° roving + ContSM)

UM-1810 (0° roving + ContSM)

UM-1810 (0° roving + ContSM)

CM-3205 (0°/90° SF + ContSM)

Bonding Layer (ChopSM)

Interior Face

FIGURE 2.3
Lay-up of face laminates.

TABLE 2.2

Layer Properties of Face Laminates

Ply Name	Ply Type	Nominal Weight, w, g/m² (oz/ft²)	Thickness, t, mm (in.)		V_f
Bonding Layer	ChSM	915.5 (3.0)	2.08	(0.082)	0.1726
CM3205	0°	542.5 (16/9)	0.62	(0.0245)	0.3428
	90°	542.5 (16/9)	0.62	(0.0245)	0.3428
	CSM	152.6 (0.5)	0.254	(0.01)	0.2359
UM1810	0°	610.3 (2.0)	0.635	(0.025)	0.3774
	CSM	305.2 (1.0)	0.335	(0.0132)	0.3582

face laminate may exhibit some extensional-bending coupling effect due to the presence of a ChSM bonding layer. In this study, the fiber system of the face laminate (see Figure 2.3) includes two layers of specified bi-ply combination mat (CM-3205) consisting of a 0°/90° SF and a CSM layer, six layers of unidirectional combination mat (UM-1810) consisting of a unidirectional layer and a CSM layer, and one bonding layer (ChSM). The details for these materials are given in Table 2.2. The resin used is polyester (UN1866).

2.1.2.2.2 Fiber Volume Fraction (V_f)

The stiffness properties of composite materials depend on the relative volume of fiber (V_f) and matrix used. For a fiber mat with nominal weight (ω), V_f can be determined from

$$V_f = \frac{\omega}{\rho \cdot t} \tag{2.2}$$

where t is the thickness of the layer, and ρ is the density of the E-glass fibers. For the face laminate given in Figure 2.3, the fiber volume fraction for each layer is computed from (2.2) and shown in Table 2.2.

2.1.2.2.3 Micromechanics

The stiffness of each ply can be predicted from micromechanics models. In this study, a micromechanics model for composites with periodic microstructure (Luciano and Barbero 1994) is used to obtain the elastic constants for each individual layer (Table 2.3). Based on the assumption that the material is isotropic in the plane, the stiffness properties of the ChSM and CSM layers are evaluated by an averaging procedure for randomly oriented composites (Harris and Barbero 1998).

2.1.2.2.4 Equivalent Properties of Face Laminate

After the elastic properties of each ply are obtained from micromechanics, the equivalent stiffness properties of the face laminate are computed from classical lamination theory (Jones 1999). A set of equivalent laminate

TABLE 2.3

Layer Stiffness Properties Obtained from Micromechanics Model

Ply Name	Orientation	E_1 (GPa)	E_2 (GPa)	G_{12} (GPa)	G_{23} (GPa)	v_{12}	v_{23}
Bond layer	Random	9.72	9.72	3.50	2.12	0.394	0.401
CM3205	0° or 90°	27.72	8.00	3.08	2.88	0.295	0.390
	Random	11.79	11.79	4.21	2.36	0.402	0.400
UM1810	0°	30.06	8.55	3.30	3.08	0.293	0.386
	Random	15.93	15.93	5.65	2.96	0.409	0.388
Core mat	Random	11.79	11.79	4.21	2.97	0.402	0.388

properties (E_x, E_y, G_{xy}, v_{xy}) can be defined for approximately balanced symmetric laminates (Davalos et al. 1996). These elastic constants represent the stiffness of an equivalent, orthotropic plate that behaves like the actual laminate under in-plane loads. The equivalent moduli of the laminate are

$$E_x = \frac{A_{11}A_{22} - A_{12}^2}{tA_{22}}, E_y = \frac{A_{11}A_{22} - A_{12}^2}{tA_{11}}, G_{xy} = \frac{A_{66}}{t}, v_{xy} = \frac{A_{12}}{A_{22}} \tag{2.3}$$

where t is the thickness of the laminated face panel and A_{ij} ($i, j = 1,2,6$) are the extensional stiffness coefficients. Based on (2.3), the equivalent properties of the face laminate are given in Table 2.4.

2.1.2.3 Modeling of Honeycomb Core

The elastic equivalence analysis of a sinusoidal-waved honeycomb core structure (see Figure 2.2) is based on a homogenization concept by a combined energy method and mechanics of materials approach. The homogenization process of periodic structures requires defining a representative volume element (RVE) or unit cell, for which the global properties can be obtained by periodical conditions and kinematical assumptions. Two different scales are defined in the homogenization process: one at the RVE level, where the mechanical properties vary from point to point within the local scale, and another at the global level, where the average properties vary smoothly over the structure. Periodical homogeneity implies that global averages and RVE averages are the same. Four fundamental concepts are applied in the analysis

TABLE 2.4

Stiffness Properties of Face Laminates

	E_x GPa (×10⁶ psi)	E_y GPa (×10⁶ psi)	v_{xy}	G_{xy} GPa (×10⁶ psi)
Face laminate	19.62 (2.846)	12.76 (1.850)	0.302	3.76 (0.546)

(Germain et al. 1983): (1) equilibrium equations, (2) the relation of averaging, (3) boundary conditions, and (4) local constitutive law.

For most sandwich cores, the RVE with regular spatial character in double symmetry gives orthotropic properties that can be defined by nine equivalent stiffness constants. The structure of the sandwich core can be separated into a number of substructures. For instance, the sinusoidal core (Figure 1.1) contains the substructures of flat and curve walls, and the walls can be simplified as series of simply supported elements without interaction with the top and bottom face laminates.

The minimum energy theory states that the strain energy calculated from the exact displacement distribution is a minimum. For a given simplified sandwich core, the averaging principle of the RVE technique can be generally expressed in parallel and series models according to Voigt and Reuss (see Christensen 1991):

$$\frac{1}{2}\frac{\sigma_{ij}^2}{C_{ij}}V \le \sum_{k=1}^{n}(U_b + U_s + U_a)_k \tag{2.4}$$

$$\frac{1}{2}C_{ij}\varepsilon_{ij}^2 V \le \sum_{k=1}^{n}(U_b + U_s + U_a)_k \tag{2.5}$$

where k accounts for individual substructures in the RVE (Figure 2.2), and U_b, U_s, and U_a are, respectively, the strain energies related to bending, shear, and axial responses. Equations (2.4) and (2.5) define, respectively, the conditions of lower and upper bounds for stiffness constants. In this study, only the energy due to bending, shear, and axial forces is included in the computation, which satisfies most periodic structures.

To obtain the elastic constants in (2.4) and (2.5), several types of loading arrangements for the lower or upper bound condition are applied to the RVE and produce a system of linear equations. The most obvious and simple loading arrangement is the application of each single principal stress or strain to obtain the corresponding stiffness without other types of strain energy involved.

When a principal load/strain is applied, the strain energy in (2.4) and (2.5) can be written as

$$U = \sum_{k=1}^{n}\left\{\int_0^s (\frac{\delta_{11}M_{11}^2}{2} + \frac{\alpha_{11}N_{11}^2}{2} + \frac{h_{44}V_{12}^2}{2})ds\right\}_k \tag{2.6}$$

where M_{11}, N_{11}, and V_{12} are the bending moment, axial force, and transverse shear force acting on the core wall (Figure 2.4), and δ_{11}, α_{11}, and h_{44} are the corresponding compliance coefficients. The subscripts 1, 2, and 3 denote local

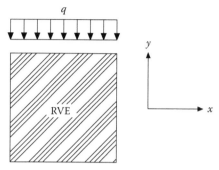

(a) RVE under stress in y direction

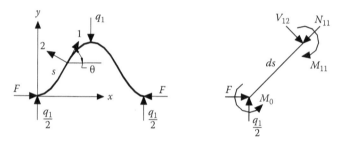

(b) Coordinate and equilibrium condition

FIGURE 2.4
Computation of Young's modulus in y direction (E_y^e).

Cartesian coordinates, with the one-axis tangent and the three-axis normal to the wave curve (see Figure 2.4).

For simplicity, there are several assumptions made in equivalence analysis of a sandwich core: the material behaves linear elastically; perfect bonding exists at face-to-core and core wall-to-wall contacts; and the ratio of thickness of the core wall to the radius of the core wall is small, and the beam bending theory can be applied. In the following section, the core equivalent properties are derived.

2.1.2.3.1 *Core Longitudinal and Transverse Young's Moduli (E_x^e and E_y^e)*

In this study, a basic element of honeycomb (RVE) is taken (Figure 2.2), and the wave function is defined by (2.1). The properties of CSM core wall can be obtained from micromechanics and are given in Table 2.3.

By applying a uniform stress q in the y direction (Figure 2.4), the internal energy U of the RVE in (2.6) becomes

$$U = 4t_2 \int_0^s (\frac{\delta_{11}M_{11}^2}{2} + \frac{\alpha_{11}N_{11}^2}{2} + \frac{h_{44}V_{12}^2}{2})ds + \frac{\alpha_{11}F^2l}{2} \tag{2.7}$$

where the bending moment, axial force, and shear force can be simply obtained by the equilibrium condition and geometric property as (see Figure 2.4)

$$M_{11} = \frac{lx}{2}q - h[1 - \cos(\frac{2\pi}{l}x)]F - M_0; \quad M_o = l^2q/8 - hF$$

$$N_{11} = \frac{h\pi\sin(\frac{2\pi}{l}x)q + F}{\sqrt{1 + \frac{\pi^2h^2}{l^2}\sin^2(\frac{2\pi}{l}x)}}; \quad V_{12} = \frac{\frac{ql}{2} - \frac{2\pi h}{l}\sin(\frac{2\pi}{l}x)F}{\sqrt{1 + \frac{\pi^2h^2}{l^2}\sin^2(\frac{2\pi}{l}x)}} \tag{2.8}$$

and for CSM core wall, the compliance coefficients are

$$\delta_{11} = \frac{12}{E_1t_2^3}; \quad \alpha_{11} = \frac{1}{E_1t_2}; \quad h_{44} = \frac{1}{\kappa G_{13}t_2} \tag{2.9}$$

where $H = 4h + 2t_1 + 2t_2$ (Figure 2.2), s = one-half curve length of a sinusoidal period starting from the origin of Cartesian coordinates, t_1 = thickness of flat core wall, t_2 = thickness of corrugated core wall, F = internal force, q = applied uniform stress parallel to the y axis, and κ = shear correction factor, 5/6 for a rectangular cross section.

The apparent strain value of RVE, ε_y, can be calculated through Castigliano's second theorem. The theorem states that under the principle of superposition, a partial derivative of the strain energy with respect to an external force gives the displacement corresponding to that force. In this case, the displacement (Δ_y) in the y direction can be obtained as

$$\Delta_y = H\varepsilon_y = \frac{\partial U}{\partial(ql)} \tag{2.10}$$

Before (2.10) is applied, the unknown internal force F can be found by imposing the compatibility condition:

$$\Delta_x = l\varepsilon_x = \frac{\partial U'}{\partial F} = \frac{Fl}{E_1t_1} \tag{2.11}$$

$$U' = 2t_2 \int_0^s (\frac{\delta_{11}M_{11}^2}{2} + \frac{\alpha_{11}N_{11}^2}{2} + \frac{h_{44}V_{12}^2}{2})ds \tag{2.12}$$

where U' is the strain energy stored in one period of sinusoidal wave.

Once the strains are defined, the equivalent stiffness (E_y^e) in the y direction and the Poisson's ratio (v_{yx}^e) can be obtained as

$$E_y^e = \frac{q}{\varepsilon_y}; \quad v_{yx}^e = -\frac{\varepsilon_x}{\varepsilon_y} = \frac{\Delta_x H}{\Delta_y l} \tag{2.13}$$

The above results based on the mechanics of materials approach lead to an exact solution, and the upper and lower bounds given in (2.4) and (2.5) result in the same solutions.

Similarly, the equivalent stiffness in the x direction can be obtained by applying a uniform stress in that direction. Based on the above formulation, the analysis indicates that the stiffness contribution of the curve beams is negligible in the x direction, and the equivalent stiffness, E_x^e, can be approximated as

$$E_x^e = \frac{2t_1}{H} E_1 \tag{2.14}$$

2.1.2.3.2 Core Out-of-Plane Shear Moduli (G_{xz}^e and G_{yz}^e)

When a shear force is applied, the induced deformation at each core wall is due to the spatial nonuniformity of the honeycomb structure. Unlike the cases of Young's moduli, the shear stiffness usually involves a complicated state of deformation, and it is relatively difficult to get an exact analytical solution. However, using the energy method, the lower and upper bound solutions can be obtained, and the lower bound usually provides a conservative solution in practical design.

When a shear stress τ_{xz} is applied in the x direction, the resulting apparent distributed shear flow is as shown in Figure 2.5. The equilibrium equation and compatibility condition are written as

$$4t_2 \int_0^s \tau_2 \cos\theta\, ds + 2t_1 l \tau_1 = Hl\tau_{xz} \tag{2.15}$$

$$2 \int_0^s \frac{\tau_2}{G_{12}^s} ds = \frac{\tau_1}{G_{12}^s} l \tag{2.16}$$

where

$$2 \int_0^s \cos\theta\, ds = 1.$$

Equations (2.15) and (2.16) lead to

$$\tau_1 = \frac{H \int_0^s ds}{l t_2 + 2t_1 \int_0^s ds} \tau_{xz} \tag{2.17}$$

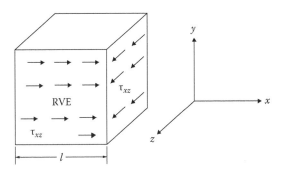

(a) Out-of-plane shear on RVE

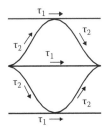

(b) Shear flow on core (plan view)

FIGURE 2.5
Computation of out-of-plane shear modulus ($G_{xz}{}^e$).

$$\tau_2 = \frac{H}{2t_2 + 4t_1 \int_0^s ds / l} \tau_{xz} \tag{2.18}$$

By applying (2.4), we can obtain:

$$\frac{G_{xz}^e}{G_{12}^s} \geq \frac{2t_1}{H} + \frac{t_2 l}{H \int_0^s ds} \tag{2.19}$$

where $G_{12}{}^s$ is the in-plane shear modulus of core wall (see Table 2.3). Similarly, applying a shear strain γ_{xz} and combining with the corresponding compatibility conditions, we can express (2.5) as

$$2t_1 l \frac{1}{2} \gamma_{xz}^2 G_{12}^s + 4 \int_0^s \frac{1}{2} G_{12}^s (\gamma_{xz} \cos\theta)^2 t_2 ds \geq \frac{Hl}{2} G_{xz}^e \gamma_{xz}^2 \tag{2.20}$$

Simplifying (2.20) gives

$$\frac{G_{xz}^e}{G_{12}^s} \leq \frac{2t_1}{H} + \frac{4t_2}{Hl} \int_0^s \cos^2\theta ds \qquad (2.21)$$

Following a similar analysis process, the out-of-plane shear modulus, G_{yz}^e, can be obtained as

$$\frac{4t_2 \int_0^s \sin^2\theta ds}{lH} G_{12}^s \geq G_{yz}^e \geq \frac{16t_2 h^2}{lH \int_0^s ds} G_{12}^s \qquad (2.22)$$

2.1.2.3.3 Summary of Core Equivalent Properties

Based on the formulations in Sections 2.1.2.3.1 and 2.1.2.3.2, the equivalent properties of FRP honeycomb core are computed and given in Table 2.5. In a similar fashion and based on a mechanics of materials approach, the Young's modulus (E_z^e) in the z direction and in-plane shear modulus (G_{xy}^e) are derived. As shown in Table 2.5, the lower and upper bounds for the out-of-plane shear moduli are within a relatively narrow range.

2.1.2.4 Equivalent Stiffness Properties for Honeycomb Sandwich Panels

By modeling a honeycomb sandwich panel as a three-layer laminated system of top and bottom faces and core and using classical lamination theory, the sandwich panel equivalent stiffness properties are computed and given in Table 2.6. The simulated layer properties for face panels and core are based on the results given in Tables 2.4 and 2.5, and for core properties, the values of lower bounds are used in the calculation.

2.1.3 Behavior of FRP Honeycomb Beams

To verify the accuracy of the equivalent stiffness properties of the FRP honeycomb panel obtained in Section 2.1.2, the experimental testing and finite element modeling of FRP honeycomb beams are performed and correlated with analytical solutions, which are based on first-order shear deformation theory. The results and comparisons for two simple loading types—three-point bending and four-point bending—are presented.

2.1.3.1 Analytical Model: Beam Theory

For a sandwich beam with thin faces, it is reasonable to assume a constant transverse shear strain through the thickness. Using Timoshenko beam theory, the coupled governing differential equations are given as

TABLE 2.5

Elastic Equivalent Properties for Honeycomb Core

Geometric data: $h = 25.4$ mm (1 in.), $l = 101.6$ mm (4 in.), $t_1 = t_2 = 2.28$ mm (0.0898 in.)

Solid core wall properties: $E_1^s = E_2^s = 11.79$ GPa (1.71×10^6 psi), $G_{12}^s = 4.21$ GPa (0.611×10^6 psi)

	E_x^e	E_y^e	E_z^e	G_{xy}^e	G_{xz}^e	G_{yz}^e	ν_{xy}^e	G_{xz}^e	G_{yz}^e
Upper bound	$0.0449\,E_1^s$	$8.36 \times 10^{-5}\,E_1^s$	$0.107\,E_2^s$	$5.98 \times 10^{-5}\,E_1^s$	$0.078\,G_{12}^s$	$0.033\,G_{12}^s$	0.431	0.169	0.273×10^{-4}
Lower bound	$0.0449\,E_1^s$	$8.36 \times 10^{-5}\,E_1^s$	$0.107\,E_2^s$	$5.98 \times 10^{-5}\,E_1^s$	$0.075\,G_{12}^s$	$0.027\,G_{12}^s$	0.431	0.169	0.273×10^{-4}

Note: e refers to quantities with equivalent properties; s refers to quantities with solid core properties.

TABLE 2.6

Equivalent Properties of FRP Honeycomb Sandwich Panel

	E_x^p GPa ($\times 10^6$ psi)	E_y^p GPa ($\times 10^6$ psi)	G_{xy}^p GPa ($\times 10^6$ psi)	ν_{xy}^p
In-plane	3.813 (0.553)	2.206 (0.320)	0.648 (0.094)	0.303
Bending	8.777 (1.273)	5.537 (0.803)	1.627 (0.236)	0.301

Note: p refers to quantities with panel properties.

$$D\frac{d^2\psi}{dx^2} + (\kappa GA)(\frac{dw}{dx} - \psi) = 0 \tag{2.23}$$

$$\kappa GA(\frac{d\psi}{dx} - \frac{d^2w}{dx^2}) = q(x) \tag{2.24}$$

where w denotes deflection, and ψ physically represents the first derivative of deflection due to bending. Beam bending stiffness D mainly consists of face in-plane tensile stiffness and core bending stiffness (Allen 1969) and is expressed as

$$D = b\left[\frac{(d-t)^2 t}{2}E_f + \frac{(d-2t)^3}{12}E_c\right] \tag{2.25}$$

where b, d, and t denote beam width, and beam depth and face thickness, respectively. The shear stiffness (κGA) of the FRP honeycomb sandwich beam can be simplified by neglecting face shear deformation, and it is expressed as

$$(\kappa GA) = G_{xz}^e bd \tag{2.26}$$

where κ is the shear correction factor and is assumed to be 1.0 for this study. By solving (2.23) and (2.24), the maximum mid-span deflections for three-point bending and four-point bending are

$$\delta_3 = \frac{PL^3}{48D} + \frac{PL}{4(\kappa GA)}, \text{ 3-point bending}$$

$$\delta_4 = \frac{23PL^3}{1296D} + \frac{PL}{6(\kappa GA)}, \text{ 4-point bending} \tag{2.27}$$

where P is the total applied load, and L is the span length. The analytical predictions for sandwich beams with sinusoidal core oriented in the longitudinal and also transverse directions are given in Tables 2.7 to 2.10. It should be noted that the assumption of thin faces is used in this study, and for thick faces, the modifications can follow the approach by Allen (1969).

TABLE 2.7

Strains and Deflections of Longitudinal Honeycomb Beams

| | | Strain, ε (με) | | | Deflection, δ (mm) | | | | | |
| | | @ 1/2 Span | | | @ 1/2 Span | | | @ 1/3 Span | | |
Span (ft)	Load Type	FE	Analysis	Experiment[a]	FE	Analysis	Experiment	FE	Analysis	Experiment
15.0	3-Point	810	1024	835, −821	26.213	28.702	25.451	22.174	24.333	20.904
	4-Point	619	683	558, −583	22.174	24.333	21.234	19.355	21.209	18.034
11.5	3-Point	700	785	633, −614	12.268	13.132	11.836	10.160	11.100	9.804
	4-Point	480	524	425, −448	10.160	11.100	9.779	8.915	9.677	8.407
8.0	3-Point	471	546	439, −401	4.343	4.572	4.089	3.607	3.835	3.404
	4-Point	330	364	295, −311	3.607	3.835	3.454	3.175	3.378	2.997
5.5	3-Point	350	376	298, −259	1.600	1.600	1.549	1.295	1.321	1.321
	4-Point	228	250	—	1.295	1.321	—	1.219	1.168	—

[a] The first value refers to the reading on the bottom of the beam (tensile), while the second one refers to the reading on the top (compressive).

Note: Beam width $d = 0.2$ m (8 in.).

TABLE 2.8

Strains and Deflections of Longitudinal Honeycomb Beams

Span (ft)	Load Type	Strain, ε ($\mu\varepsilon$) @ 1/2 Span			Deflection, δ (mm) @ 1/2 Span			@ 1/3 Span		
		FE	Analysis	Experiment[a]	FE	Analysis	Experiment	FE	Analysis	Experiment
15.0	3-Point	620	683	570, −522	17.450	19.126	17.374	14.757	16.231	14.148
	4-Point	361	455	374, −378	14.757	16.231	14.529	12.878	14.148	12.167
11.5	3-Point	440	524	438, −388	8.026	8.738	8.153	6.756	7.391	6.655
	4-Point	325	349	288, −291	6.756	7.391	6.680	5.918	6.452	5.690
8.0	3-Point	330	364	303, −252	2.870	3.048	2.921	2.388	2.565	2.337
	4-Point	220	243	200, −204	2.388	2.565	2.464	2.108	2.235	2.083
5.5	3-Point	210	250	209, −159	1.067	1.067	1.295	0.864	0.889	1.016
	4-Point	155	167	—	0.864	0.889	—	0.787	0.787	—

[a] The first value refers to the reading on the bottom of the beam (tensile), while the second one refers to the reading on the top (compressive).

Note: Beam width $d = 0.3$ m (12 in.).

TABLE 2.9

Strains and Deflections of Transverse Honeycomb Beams

| | | Strain, ε (με) | | | Deflection, δ (mm) | | | | | |
| | | @ 1/2 Span | | | @ 1/2 Span | | | @ 1/3 Span | | |
Span (ft)	Load Type	FE	Analysis	Experiment[a]	FE	Analysis	Experiment	FE	Analysis	Experiment
8.0	3-Point	860	892	881, −709	7.772	7.722	8.509	6.426	6.426	6.909
	4-Point	572	595	570, −559	6.426	6.426	6.985	5.690	5.664	6.020
5.5	3-Point	599	614	590, −442	2.946	2.769	3.073	2.413	2.261	2.540
	4-Point	399	409	—	2.413	2.261	—	2.159	2.007	—

[a] The first value refers to the reading on the bottom of the beam (tensile), while the second one refers to the reading on the top (compressive).
Note: Beam width $d = 0.2$ m (8 in.).

TABLE 2.10

Strains and Deflections of Transverse Honeycomb Beams

| | | Strain, ε (με) | | | Deflection, δ (mm) | | | | | |
| | | @ 1/2 Span | | | @ 1/2 Span | | | @ 1/3 Span | | |
Span (ft)	Load Type	FE	Analysis	Experiment[a]	FE	Analysis	Experiment	FE	Analysis	Experiment
8.0	3-Point	621	595	564, −500	5.156	5.131	5.258	4.267	4.293	4.318
	4-Point	416	397	382, −395	4.267	4.293	4.470	3.835	3.759	3.810
5.5	3-Point	419	409	386, −306	1.956	1.854	2.032	1.600	1.524	1.676
	4-Point	270	273	—	1.600	1.524	—	1.422	1.346	—

[a] The first value refers to the reading on the bottom of the beam (tensile), while the second one refers to the reading on the top (compressive).
Note: Beam width $d = 0.3$ m (12 in.).

2.1.3.2 Finite Element Modeling

The finite element (FE) program ANSYS (version 5.5) was used to model the actual honeycomb sandwich beams. The face laminate consists of triangular six-node isoparametric shell elements that allow input of up to 256 composite layers. The sinusoidal and flat laminates of the core wall are modeled with rectangular eight-node isoparametric shell elements, and the sinusoidal wave shape of the core wall is defined by a polynomial function. The global size of each element is set at about 1 in., and for example, there are more than 3,000 elements for the 8 in. wide and 8 ft long beam. The numerical predictions are given in Tables 2.7 to 2.10 for various beams and load conditions. The experimental results and their correlations with finite element analyses and analytical predictions are presented next.

2.1.3.3 Experimental Study

To indirectly verify the accuracy of the equivalent orthotropic properties given in Table 2.6, which are obtained from homogenization analysis of core and micro/macromechanics analysis of face laminates, the longitudinal and transverse FRP honeycomb beams were tested. Two beam widths of 0.2032 m (8.0 in.) and 0.3048 m (12.0 in.) were provided by KSCI, and two beams for each beam type were tested under bending. At the mid-span for each sample, three strain gages were bonded along the width of the top face and one at the center of the bottom face. Displacement transducers (Linear Variable Differential Transducers, LVDTs) were also placed at the center and one-third spans. The beams were tested under both three-point and four-point bending loads. In the three-point bending test, the load was applied at the mid-span, whereas for the four-point bending, the loads were applied at the third and two-third points of the span. To study the shear effect on deflection, various span lengths were tested, and the experimental results are summarized in Tables 2.7 to 2.10.

2.1.3.4 Comparison and Summary

As indicated in Tables 2.7 to 2.10, relatively good agreements among the experimental data, FE predictions, and analytical results are achieved. Especially, the deflections for both the longitudinal and transverse beams based on analytical predictions compare well with the experimental and finite element model results.

2.1.4 Behavior of FRP Honeycomb Sandwich Panel

As an application, a full-size FRP honeycomb panel of $2.36 \times 4.57 \times 1.52$ m (7.75 ft × 15 ft × 5 in.) (Figure 2.6) is tested under bending and also analyzed by the finite element method. The panel deck is simply supported over a span of 4.57 m (15 ft) and subjected to a patch load (0.2413×0.6096 m) (9.5×24 in.).

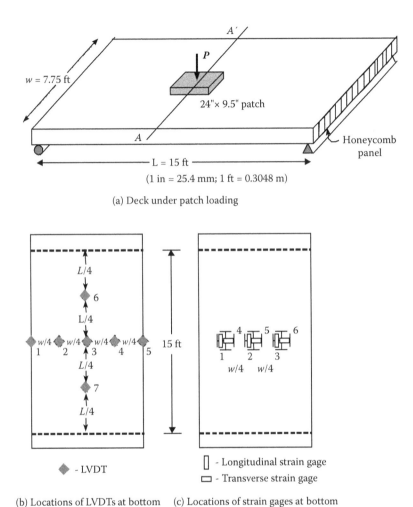

FIGURE 2.6
Experimental setup of FRP honeycomb deck panel.

Three load conditions are applied at mid-span to simulate symmetric and asymmetric cases (see Figure 2.6): (a) at the center of the deck, (b) at one-fourth ($w/4$) of the width from one edge along mid-span section A-A, and (c) at $w/4$ from the opposite edge along mid-span section A-A. For each load condition, the displacements were recorded at seven locations with LVDTs (see Figure 2.6(b)), and the strains in both longitudinal and transverse directions were obtained at three locations along the mid-span using bonded strain gages at the bottom of the deck (Figure 2.6(c)). From Figure 2.6, note that for the asymmetric load cases 2 and 3, the following displacement values should correspond to each other: δ_1 and δ_5, δ_2 and δ_4, and δ_6 and δ_7, since they are symmetric with respect to the center of the deck. Similarly, the following strains should correspond to each other: ε_1 and ε_3, and ε_4 and ε_6.

In the finite element analysis, a sandwich panel with equivalent three layers (top and bottom faces and core) is modeled. For simplicity and verification purposes, the equivalent properties obtained for face laminates (Table 2.4) and core (Table 2.5) are used directly in the model, and the face laminates and core are each modeled as a single layer using eight-node shell elements. The mesh used with and the displacement contours obtained from ANSYS FE are shown in Figure 2.7. As shown in Table 2.11, the finite element predictions based on equivalent material properties compare favorably with experimental data. For the symmetric loading, the maximum difference for deflection is 8.56%, while for asymmetric cases, the maximum difference is about 11.62%.

2.1.5 Conclusions

In summary, this section presents a combined analytical and experimental characterization of FRP honeycomb panels. The core consists of in-plane sinusoidal cells extending vertically between top and bottom face laminates. The emphasis of this study is on evaluation of equivalent properties for both face and core components. A combined micro/macromechanics approach is used to predict face laminate elastic properties, and the core equivalent properties are obtained by a homogenization technique combined with an energy method and a mechanics of materials approach. The analytical model predictions correlate well with finite element modeling and experimental results for both beam and plate specimens. In particular, the test results for a 2.36×4.57 m (7.75×15 ft) panel under symmetric and asymmetric patch loads match well with finite element predictions obtained by modeling the specimen as a three-layer equivalent plate. Thus, the equivalent orthotropic properties developed in this section can be used with confidence in design analyses of the FRP sandwich panel.

2.2 On the Transverse Shear Stiffness of Composite Honeycomb Core with General Configuration

In this section, the effective transverse shear stiffness properties of a generic composite honeycomb core configuration based on a homogenization technique (Xu et al. 2001) are presented.

2.2.1 Introduction

By observing the repetitive pattern of sandwich structures, it is practicable to approach the solution as a boundary value problem with effective stiffness

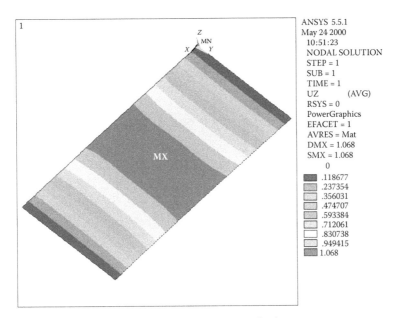

(a) Panel displacement contour under symmetric loading

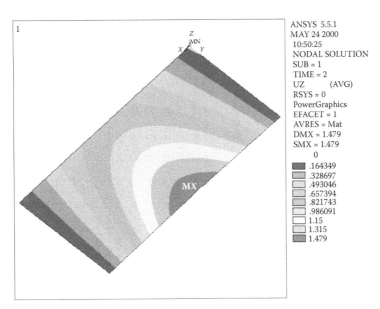

(b) Panel displacement contour under asymmetric loading

FIGURE 2.7
FRP sandwich panel (ANSYS).

TABLE 2.11

Experimental and Finite Element Comparison for a FRP Honeycomb Deck

Load Type		Deflection (mm)							Strain (με)					
		δ_1	δ_2	δ_3	δ_4	δ_5	δ_6	δ_7	ε_1	ε_2	ε_3	ε_4	ε_5	ε_6
Case 1	Experiment	24.638	24.790	25.400	24.689	24.714	20.803	21.158	825	855	826	-113	283	-48
	FE	26.746	26.619	27.127	26.619	26.746	22.428	22.428	1003	1092	1003	-175	168	-175
Case 2	Experiment	33.655	29.235	24.765	20.803	17.983	20.701	21.184	1069	799	667	198	-120	-161
	FE	37.567	32.207	26.670	22.047	18.542	22.555	22.555	1357	1033	819	128	-236	-170
Case 3	Experiment	17.856	20.828	25.146	29.845	34.442	20.701	21.234	664	844	1019	-189	-112	103
	FE	18.542	22.047	26.670	32.207	37.567	22.555	22.555	819	1033	1357	-170	-236	128

properties by treating a spatially heterogeneous structure as an equivalent homogeneous bulk material. The effective stiffness may be analytically predicted when the local problem on a characterized representative volume element (RVE) or unit cell is reduced to a two-dimension case, and in general, thin-walled honeycomb core configurations with warping-free assumption may satisfy this condition. Since previous research was mainly focused on the traditional hexagonal honeycomb core made of metallic foil, there is little literature available in the sense of general spatial configurations. With regard to computational models on sandwich structures, Noor et al. (1996) presented a review wherein most of relevant literature is listed. The early research in hexagonal honeycomb includes the work of Kelsey et al. (1958), where upper and lower bounds of transverse shear stiffness were obtained with the classical energy method. Among others, Gibson and Ashby (1988) presented the predictions for transverse shear stiffness of hexagonal honeycomb using mechanics of materials and energy methods. Hohe et al. (1999) studied general hexagonal and quadrilateral grid structures by using an energy approach. All these classical mechanics approaches are effective for specific problems of isotropic grid honeycomb, but they become somewhat limited when applied to honeycomb problems with general shapes and anisotropic materials. In general, this limitation is partially related to conventional methods themselves, which lack rigor of mathematical theory.

The mathematical description of micro-periodical composite materials has been well developed during the 1970s. The method of homogenization is believed capable of approximating the equivalent stiffness properties effectively and unrivaled by any other known methods in terms of accuracy and closeness. When applying homogenization techniques at micro and macro levels, there are basically no distinctions between problems of inhomogeneous materials and discrete network structures. The homogenization method for composite materials (Meguid and Kalamkarov 1994) can thus be analogically used in structural problems of thin-walled honeycomb cores. However, the application of the homogenization method to engineering problems is not easy, and little work has been done in this area. A crucial issue is the solution of a special local problem assisted by physical interpretation of the localized variables, by means of which the homogenization process can thus be less cumbersome and more expedient. In the field of sandwich structures, a good attempt was made by Shi and Tong (1995a) in presenting an analytical solution for a hexagonal honeycomb core using the homogenization theory.

Of all the effective stiffness properties of the honeycomb sandwich core, the transverse shear properties are most difficult to predict. Also, due to the relatively low shear moduli of polymer resins in composite materials, an accurate prediction of effective transverse shear stiffness of a composite honeycomb core becomes more important. In this section, the transverse shear stiffness of a honeycomb core with a general configuration is investigated, and a two-scale homogenization method is used to obtain explicit formulas for general configurations of thin-walled honeycomb structures. The analytical lower bound formula of effective transverse shear stiffness is formulated by means of the homogenization method. The approach adopts basic mathematical concepts of homogenization theory, and the 3D partial differential equations are solved with the assumption of free warping and constant variables through core wall thickness. This approach can be further extended to other honeycomb core structures, by implementing certain modifications when required; e.g., the effect of core wall thickness can be added for thick core wall structures. The basic mechanics concepts of the homogenization method are used in this section, and details can be found in a number of relevant references (Kalamkarov 1992; Parton and Kudryavtsev 1993). Several practical examples of honeycomb cores with different configurations are solved with the derived formula, which is validated by existing approaches (e.g., the solutions given by Kelsey et al. (1958) and Shi and Tong (1995a) for hexagonal honeycomb core) or is verified by finite element analyses presented in this section.

2.2.2 Application of Homogenization Theory

Honeycomb sandwich structures usually consist of two outer facesheets and a honeycomb core with double periodicity in the plane normal to the thickness direction. In the homogenization process, a parallelepiped unit cell is first defined and selected to characterize the spatial periodicity of a sandwich structure. We consider a body of honeycomb sandwich (Figure 2.8) occupying a bound region Ω in R^3 space, defined by coordinates x_1, x_2, and x_3, with a smooth boundary under body force P_i. The region Ω consists of a doubly periodic unit cell Y with in-plane dimensions of εy_1, εy_2, and thickness y_3

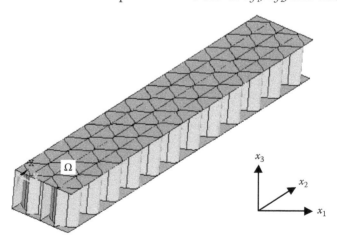

FIGURE 2.8
A body of honeycomb sandwich structure.

in the same order. It should be pointed out that the homogenization of a
3D periodicity body is different from that of plates with a thickness dimension much smaller than that of the other two. However, when we neglect the
warping constraints of a sandwich facesheet, the estimate of transverse shear
stiffness can be considered independent of thickness dimension. Therefore,
we can rescale the thickness dimension to attain the same periodicity parameter ε, by which the 3D asymptotic expansion (Parton and Kudryavtsev 1993)
can be simply adopted in the following derivations, wherein the notation is
given with small Latin indices $h, i, j, k, l = 1, 2, 3$ and small Greek indices $\alpha,$
$\beta, \lambda = 1,2$. The equations of equilibrium and the boundary conditions of the
linear elasticity problem may be written as

$$\frac{\partial \sigma_{ij}^{(\varepsilon)}}{\partial x_j} + P_i = 0, \quad \text{in } \Omega$$

$$u_i^{(\varepsilon)} = \hat{u}_i, \quad \text{on } \partial_1 \Omega \tag{2.28}$$

$$\sigma_{ij}^{(\varepsilon)} n_j = T_i, \quad \text{on } \partial_2 \Omega$$

where

$$\sigma_{ij}^{(\varepsilon)} = c_{ijkl}\left(\frac{x}{\varepsilon}\right)e_{kl}^{(\varepsilon)}$$

$$e_{kl}^{(\varepsilon)} = \frac{1}{2}\left(\frac{\partial u_k^{(\varepsilon)}}{\partial x_l} + \frac{\partial u_l^{(\varepsilon)}}{\partial x_k}\right) \tag{2.29}$$

And the coefficient $c_{ijkl}(y)$ should satisfy the elliptical symmetry condition

$$c_{ijkl}(y) = c_{jikl}(y) = c_{ijlk}(y) = c_{klij}(y) \tag{2.30}$$

By using the two-scale expansion method, the series are expressed as

$$u_i^{(\varepsilon)} = u_i^{(0)}(x) + \varepsilon\, u_i^{(1)}(x,y) + \cdots$$

$$e_{ij}^{(\varepsilon)} = e_{ij}^{(0)}(x,y) + \varepsilon\, e_{ij}^{(1)}(x,y) + \cdots \tag{2.31}$$

$$\sigma_{ij}^{(\varepsilon)} = \sigma_{ij}^{(0)}(x,y) + \varepsilon\, \sigma_{ij}^{(1)}(x,y) + \cdots$$

where

$$e_{ij}^{(0)}(x,y) = \frac{1}{2}(\frac{\partial u_i^{(0)}}{\partial x_j} + \frac{\partial u_j^{(0)}}{\partial x_i}) + \frac{1}{2}(\frac{\partial u_i^{(1)}}{\partial y_j} + \frac{\partial u_j^{(1)}}{\partial y_i})$$

$$e_{ij}^{(1)}(x,y) = \frac{1}{2}(\frac{\partial u_i^{(1)}}{\partial x_j} + \frac{\partial u_j^{(1)}}{\partial x_i}) + \frac{1}{2}(\frac{\partial u_i^{(2)}}{\partial y_j} + \frac{\partial u_j^{(2)}}{\partial y_i}) \tag{2.32}$$

$$\sigma_{ij}^{(0)}(x,y) = c_{ijkh}(y)\frac{\partial u_k^{(0)}}{\partial x_h} + c_{ijkh}(y)\frac{\partial u_k^{(1)}}{\partial y_h}$$

$$\sigma_{ij}^{(1)}(x,y) = c_{ijkh}(y)\frac{\partial u_k^{(1)}}{\partial x_h} + c_{ijkh}(y)\frac{\partial u_k^{(2)}}{\partial y_h} \tag{2.33}$$

It is noted that the variables with superscripts 1 and 2 are Y-periodic in y. Substituting (2.32) and (2.33) into (2.28) and retaining the terms of $O(\varepsilon^{-1})$ and $O(\varepsilon^0)$, we have

$$\frac{\partial \sigma_{ij}^{(0)}}{\partial y_j} = 0 \tag{2.34}$$

$$\frac{\partial \sigma_{ij}^{(0)}}{\partial x_j} + \frac{\partial \sigma_{ij}^{(1)}}{\partial y_j} + P_i = 0 \tag{2.35}$$

Substituting (2.33) into (2.34) gives

$$\frac{\partial}{\partial y_j}\left(c_{ijkh}(y)\frac{\partial u_k^{(1)}(x,y)}{\partial y_h} \right) = -\frac{\partial u_k^{(0)}(x)}{\partial x_l}\frac{\partial c_{ijkl}(y)}{\partial y_j} \tag{2.36}$$

In (2.36), we may introduce the Y-periodic functions by

$$u_n^{(1)}(x,y) = N_n^{kl}(y)\frac{\partial u_k^{(0)}(x)}{\partial x_l} \tag{2.37}$$

Thus, (2.36) becomes

$$\frac{\partial}{\partial y_j}\left(c_{ijnh}(y)\frac{\partial N_n^{kl}(y)}{\partial y_h} \right) = -\frac{\partial c_{ijkl}(y)}{\partial y_j} \tag{2.38}$$

Let

$$\tau_{ij}^{kl}(y) = c_{ijnh}(y)\frac{\partial N_n^{kl}(y)}{\partial y_h} \tag{2.39}$$

Then, (2.38) can be written as

$$\frac{\partial \tau_{ij}^{kl}(y)}{\partial y_j} = -\frac{\partial c_{ijkl}(y)}{\partial y_j} \tag{2.40}$$

Since the function is Y-periodic in y, (2.35) will have a unique solution if the following condition is satisfied;

$$\int_Y \frac{\partial \sigma_{ij}^{(1)}}{\partial y_j} dy = \int_{\partial Y} \sigma_{ij}^{(1)} n_j \, dS_Y = 0 \tag{2.41}$$

where Y represents a parallelepiped unit cell, and n_j is the unit vector in the outward normal direction to the surface ∂Y.

Applying a volume averaging procedure to (2.35) over Y and considering the conditions given in (2.32), (2.33), and (2.41), we have

$$\frac{\partial \left\langle \sigma_{ij}^{(0)} \right\rangle}{\partial x_j} + P_i = 0 \tag{2.42}$$

where

$$\left\langle \sigma_{ij}^{(0)} \right\rangle = \frac{1}{|Y|} \int_Y \sigma_{ij}^{(0)} \, dy = \left\langle c_{ijkl} \right\rangle \frac{\partial u_k^{(0)}(x)}{\partial x_l} \tag{2.43}$$

and

$$\left\langle c_{ijkl} \right\rangle = \frac{1}{|Y|} \int_Y \left(c_{ijkl}(y) + \tau_{ij}^{kl}(y) \right) dy \tag{2.44}$$

The effective elastic constants of a unit cell are defined by $\left\langle c_{ijkl} \right\rangle$. Based on the so-called zeroth-order approximation given in (2.28) to (2.44), the equivalent elastic coefficients can be computed with the solution of the local function $\tau_{ij}^{kl}(y)$.

2.2.3 Derivation of Effective Transverse Shear Stiffness

2.2.3.1 Description of 2D Periodic Thin-Walled Honeycomb Core

A unit cell of a general honeycomb core structure (in a coordinate system of y_1, y_2, and y_3) is shown in Figure 2.9, where the whole domain and the region of composite laminates inside are designated as Y and Y_s, respectively, and the thickness of honeycomb is denoted as δ. Within Y_s, a segment AB with arbitrary in-plane curve function is selected for analysis. Due to periodicity, the ends A and B are located at the opposite side of the unit cell. Let s and η denote two local in-plane coordinates, one tangential along the curve segment and the other in the normal direction. For simplicity, the composite core material is assumed to be orthotropic and defined by nine elastic constants, where G_L and G_T denote the in-plane and transverse shear stiffness

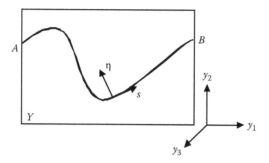

FIGURE 2.9
Unit cell element of a general honeycomb core.

properties, respectively. Obviously the material shear stiffnesses c_{1313}, c_{2323}, and c_{1323} in the y coordinate system (Figure 2.9) can be calculated by transformations of G_L and G_T.

For thin-walled structures, the thickness of core wall $t(s)$ is assumed small compared with the size of the unit cell. Therefore, it is reasonable to assume constant variables through the wall thickness. When we apply (2.44) to a honeycomb core consisting of discrete structures, the averaging integration can be made by a summation of all segments:

$$\langle c_{\alpha 3 \beta 3} \rangle = \frac{1}{|Y|} \int_Y \left(c_{\alpha 3 \beta 3}(y) + \tau_{\alpha 3}^{\beta 3} \right) dy = \frac{1}{|Y|} \left(\sum_K \left[\int_\delta \int_A^B \left(c_{\alpha 3 \beta 3}(y) + \tau_{\alpha 3}^{\beta 3} \right) t(s) \, ds \, dy_3 \right]_K \right) \quad (2.45)$$

where K itemizes the segments.

To calculate the effective shear stiffness by (2.45), we may first obtain analytical solutions for individual segments. It is observed that the transverse shear stiffnesses G_L and G_T of the unit cell material are zero in $Y \backslash Y_s$ and constant in Y_s. This situation by nature is similar to that of periodically distributed holes in a bulk material. The local boundary condition can be written as

$$\sigma^{(0)} \cdot n = 0 \quad (2.46a)$$

Applying (2.33) and (2.37), (2.46a) can be expressed as

$$\left(c_{ij\alpha 3}(y) + \tau_{ij}^{\alpha 3}(y) \right) \cdot n_j = 0, \quad \text{at } \partial Y_s \quad (2.46b)$$

where n is the unit vector normal to the core vertical wall ∂Y_s. Physically, the above boundary condition represents traction-free on the vertical surface of the core wall.

The condition of periodicity (i.e., duplication of local function $N_i^{\alpha 3}(y)$ on A and B) is expressed as

$$\left[N_i^{\alpha 3}(y) \right]_A = \left[N_i^{\alpha 3}(y) \right]_B \tag{2.47}$$

As given in Equations (2.45) to (2.47), the homogenization process actually reduces to solving the unknown local function $\tau_{ij}^{kl}(y)$ with the application of local periodic boundary conditions.

2.2.3.2 Determination of Local Function

Based on the warping-free assumption that the local variables are independent of thickness dimension y_3, (2.40) may be specifically expressed as follows:

$$\frac{\partial \tau_{13}^{\alpha 3}}{\partial y_1} + \frac{\partial \tau_{23}^{\alpha 3}}{\partial y_2} = -\left(\frac{\partial c_{13\alpha 3}}{\partial y_1} + \frac{\partial c_{23\alpha 3}}{\partial y_2} \right), \quad \text{in } Y_s \tag{2.48a}$$

$$\frac{\partial \tau_{12}^{\alpha 3}}{\partial y_2} + \frac{\partial \tau_{11}^{\alpha 3}}{\partial y_1} = 0, \quad \text{in } Y_s \tag{2.48b}$$

where

$$\tau_{13}^{\alpha 3} = c_{1313} \frac{\partial N_3^{\alpha 3}(y)}{\partial y_1} + c_{1323} \frac{\partial N_3^{\alpha 3}(y)}{\partial y_2} \tag{2.49a}$$

$$\tau_{23}^{\alpha 3} = c_{2313} \frac{\partial N_3^{\alpha 3}(y)}{\partial y_1} + c_{2323} \frac{\partial N_3^{\alpha 3}(y)}{\partial y_2} \tag{2.49b}$$

Note in (2.49) that the term

$$\frac{\partial N_\beta^{\alpha 3}(y)}{\partial y_3}$$

is taken to zero by the warping-free assumption. Further (2.48a) can be rewritten in local coordinates (s, η) as

$$\frac{\partial}{\partial s}\left(\tau_{13}^{\alpha 3} + c_{13\alpha 3} \right) \frac{ds}{dy_1} + \frac{\partial}{\partial \eta}\left(\tau_{23}^{\alpha 3} + c_{23\alpha 3} \right) \frac{d\eta}{dy_2} = 0, \quad \text{in } Y_s \tag{2.50a}$$

$$\frac{\partial}{\partial s}\left(\tau_{23}^{\alpha 3} + c_{23\alpha 3} \right) \frac{ds}{dy_2} + \frac{\partial}{\partial \eta}\left(\tau_{13}^{\alpha 3} + c_{13\alpha 3} \right) \frac{d\eta}{dy_1} = 0, \quad \text{in } Y_s \tag{2.50b}$$

As is the case for thin-walled structures, the gradient through the thickness η is approximately zero; thus the second terms of Equation (2.50) disappear, and (2.50) reduces to

$$\tau_{13}^{\alpha3}(s) + c_{13\alpha3}(s) = \cos\theta \; const^{\alpha1} \tag{2.51a}$$

$$\tau_{23}^{\alpha3}(s) + c_{23\alpha3}(s) = \sin\theta \; const^{\alpha2} \tag{2.51b}$$

where θ denotes the anticlockwise angle from y_1 to s, and $const^{\alpha1}$ and $const^{\alpha2}$ are constants.

The consideration of local boundary conditions in (2.46) results in the condition of constant shear flow and $const^{\alpha1} = const^{\alpha2} = const^{\alpha}$. Then, (2.51) becomes

$$\tau_{13}^{\alpha3}(s) + c_{13\alpha3}(s) = \cos\theta \; const^{\alpha} \tag{2.52a}$$

$$\tau_{23}^{\alpha3}(s) + c_{23\alpha3}(s) = \sin\theta \; const^{\alpha} \tag{2.52b}$$

Physically, the above equations can be interpreted as constant shear flows along a segment. To further take into account the wall thickness function $t(s)$ along the segment, (2.52) may be modified as

$$\tau_{13}^{\alpha3}(s) + c_{13\alpha3}(s) = \cos\theta \frac{const^{\alpha}}{t(s)} \tag{2.53a}$$

$$\tau_{23}^{\alpha3}(s) + c_{23\alpha3}(s) = \sin\theta \frac{const^{\alpha}}{t(s)} \tag{2.53b}$$

To determine $const^{\alpha}$ in the above expressions, we use the relations given in (2.49):

$$\tau_{13}^{\alpha3}(s) = G_L \cos\theta \frac{\partial N_3^{\alpha3}}{\partial s} - G_T \sin\theta \frac{\partial N_3^{\alpha3}}{\partial\eta} \tag{2.54a}$$

$$\tau_{23}^{\alpha3}(y) = G_L \sin\theta \frac{\partial N_3^{\alpha3}}{\partial s} + G_T \cos\theta \frac{\partial N_3^{\alpha3}}{\partial\eta} \tag{2.54b}$$

with coordinate transformations

$$c_{1313} = G_L \cos^2\theta + G_T \sin^2\theta \tag{2.55a}$$

$$c_{2323} = G_L \sin^2\theta + G_T \cos^2\theta \tag{2.55b}$$

$$c_{1323} = (G_L - G_T)\sin\theta\cos\theta \tag{2.55c}$$

Substituting (2.54) and (2.55) into (2.53) and after several steps of transformation, we can finally have

$$\frac{\partial N_3^{13}}{\partial s} = \frac{const^1}{t(s)G_L} - \cos\theta \tag{2.56a}$$

$$\frac{\partial N_3^{13}}{\partial \eta} = \sin\theta \tag{2.56b}$$

$$\frac{\partial N_3^{23}}{\partial s} = \frac{const^2}{t(s)G_L} - \sin\theta \tag{2.56c}$$

$$\frac{\partial N_3^{23}}{\partial \eta} = \cos\theta \tag{2.56d}$$

For a simple curve AB without ramification, the substitution of (2.55a) and (2.55c) into (2.47) results in

$$\int_A^B \frac{\partial N_3^{13}}{\partial s}ds = \int_A^B \left(\frac{const^1}{t(s)G_L} - \cos\theta \right)ds = 0 \tag{2.57a}$$

$$\int_A^B \frac{\partial N_3^{23}}{\partial s}ds = \int_A^B \left(\frac{const^2}{t(s)G_L} - \sin\theta \right)ds = 0 \tag{2.57b}$$

If a segment is ramified, $const^\alpha$ becomes piecewise constant, and the continuity of shear flow should be considered at joints. For a segment with ramification, (2.57) can thus be modified as

$$\int_A^B \frac{\partial N_3^{13}}{\partial s}ds = \int_A^B \left(\frac{const^{1m}}{t(s)G_L} - \cos\theta \right)ds = 0 \tag{2.58a}$$

$$\int_A^B \frac{\partial N_3^{23}}{\partial s}ds = \int_A^B \left(\frac{const^{2m}}{t(s)G_L} - \sin\theta \right)ds = 0 \tag{2.58b}$$

where m is the number of ramification joints along the segment AB, and $const^{1m}$ and $const^{2m}$ denote the piecewise constant shear flow along the segment.

Since the plane arrangement of segments in a unit cell is generally doubly symmetric, the problem of ramified segments can be simplified and is shown for the hexagonal honeycomb core of the succeeding section. For the cases of nonramification, (2.57) is directly solved as

$$const^{\alpha} = \frac{d_{\alpha}}{T} G_L \qquad (2.59)$$

where

$$d_1 = \int_A^B \cos\theta\, ds; \quad d_2 = \int_A^B \sin\theta\, ds; \quad T = \int_A^B \frac{1}{t(s)}\, ds \qquad (2.60)$$

Substituting the solution (2.59) into (2.53), we can finally obtain the local functions as

$$\tau_{13}^{\alpha 3}(s) + c_{13\alpha 3}(s) = \cos\theta \frac{d_{\alpha}}{Tt(s)} G_L \qquad (2.61a)$$

$$\tau_{23}^{\alpha 3}(s) + c_{23\alpha 3}(s) = \sin\theta \frac{d_{\alpha}}{Tt(s)} G_L \qquad (2.61b)$$

2.2.3.3 Results and Applications

By substituting the local functions (2.61) derived above into (2.45), the explicit formula of effective transverse shear stiffness can be expressed as

$$\langle c_{\alpha 3\beta 3} \rangle = \frac{1}{|Y|} \sum_K \left[\int_\delta^B \int_A (c_{\alpha 3\beta 3}(y) + \tau_{\alpha 3}^{\beta 3}) t(s)\, ds\, dy_3 \right]_K = \frac{G_L}{\Lambda} \sum_K \left[\frac{d_{\alpha} d_{\beta}}{T} \right]_K, \quad \alpha, \beta = 1, 2 \quad (2.62)$$

where

$$d_1 = \int_A^B \cos\theta\, ds; \quad d_2 = \int_A^B \sin\theta\, ds; \quad T = \int_A^B \frac{1}{t(s)}\, ds$$

and Λ denotes the area of unit cell in the plane of y_1 and y_2, and ds represents the infinitesimal length of segment K.

As described in (2.62), the contribution of each segment to the effective shear stiffness depends on the term

$$\frac{d_{\alpha} d_{\beta}}{T}.$$

If a segment is straight with constant thickness t and aligned with the coordinate y_{α}, then $d_{\beta} = 0$, $d_{\alpha} = Tt$; thus, its contributions to $\langle c_{\alpha 3\alpha 3} \rangle$, $\langle c_{\alpha 3\beta 3} \rangle$, $\langle c_{\beta 3\beta 3} \rangle$

are $d_\alpha t$, 0, 0, respectively, which are exactly identical to those familiar forms used for I-section beams in beam theory.

To demonstrate the generality of formula (2.62), three typical honeycomb core structures with different core configurations are studied and discussed next.

2.2.3.3.1 Reinforced Sinusoidal Honeycomb Core

The configuration of a sinusoidal corrugation between horizontal flat sections is shown in Figures 2.8 and 2.10. The thicknesses of the core wall segments are assumed to be constant, and they are given as t_1 and t_2 for flat and curve segments, respectively. The function of OB (see Figure 2.10(b)) is written as

$$y_2 = \frac{b}{2}(1 - \cos\frac{\pi y_1}{a}), \quad 0 \le y_1 \le \frac{a}{2} \tag{2.63}$$

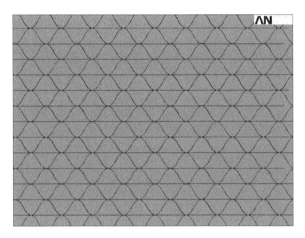

(a) Honeycomb structure

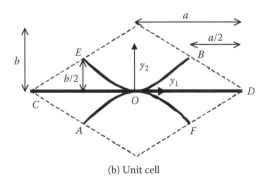

(b) Unit cell

FIGURE 2.10
Sinusoidal core: (a) honeycomb structure and (b) unit cell.

For segment AOB, it follows from (2.60) that $d_1 = a$, $d_2 = b$, and $T = S/t_2$, where S denotes the length of segment AOB. Similarly, we have $d_1 = a$, $d_2 = -b$, and $T = S/t_2$ for segment EOF, and $d_1 = 2a$, $d_2 = 0$, and $T = 2a/t_1$ for segment COD. By substituting these results and the area $\Lambda = 2ab$ into the summation of (2.62), the effective shear stiffness properties are easily calculated as

$$\langle c_{1313} \rangle = (\frac{t_1}{b} + \frac{at_2}{bS})G_L; \quad \langle c_{2323} \rangle = \frac{bt_2}{aS}G_L; \quad \langle c_{1323} \rangle = 0 \tag{2.64}$$

where S is the length for the segment,

$$S = \int_A^B ds.$$

2.2.3.3.2 Tubular Honeycomb Core

The configuration of a tubular honeycomb core is shown in Figure 2.11. The radius of curvature is equal to R. As noticed, segments AE and BF are not continuous within the unit cell; we imagine a virtual segment CD connecting them. Thus, using (2.60), we can have the calculated values of d_1, d_2, and T (see Table 2.12).

Using the area $\Lambda = 2\sqrt{3}R^2$ and (2.62), the effective shear stiffness properties for tubular honeycomb core are calculated as

$$\langle c_{1313} \rangle = \frac{(51 - 24\sqrt{3})t}{2\sqrt{3}\pi R}G_L; \quad \langle c_{2323} \rangle = \frac{9t}{2\sqrt{3}\pi R}G_L; \quad \langle c_{1323} \rangle = 0 \tag{2.65}$$

2.2.3.3.3 Hexagonal Honeycomb Core

Hexagonal honeycomb geometry is commonly used for sandwich core, for which the effective transverse shear stiffness properties were evaluated

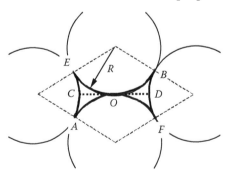

FIGURE 2.11
Tubular core.

TABLE 2.12

Constants for Tubular Honeycomb Core

	AOB	EOF	A(CD)B	E(CD)F
d_1	$\sqrt{3}R$	$\sqrt{3}R$	$(2-\sqrt{3})R$	$(2-\sqrt{3})R$
d_2	R	$-R$	R	$-R$
T	$\dfrac{2\pi R}{3t}$	$\dfrac{2\pi R}{3t}$	$\dfrac{\pi R}{3t}$	$\dfrac{\pi R}{3t}$

early by Kelsey et al. (1958). In this study, we take this configuration as an example to illustrate the simplification procedure for ramification problems. As shown in Figure 2.12, there are two joints C and D where ramifications begin. The thickness of horizontal and inclined walls is given as t_1 and t_2, respectively. For computation of $\langle c_{1313} \rangle$, segments ACDB and ECDF are simulated as two separate ones due to geometrical symmetry. Therefore, by direct application of (2.60), we have d_1 and T for the two segments as

$$d_1 = a + b\cos\theta$$

$$T = \frac{2a}{t_1} + \frac{b}{t_2} \tag{2.66}$$

Using the summation given in (2.62), we obtain

$$\langle c_{1313} \rangle = \frac{G_L}{A} \frac{2(a+b\cos\theta)^2}{\dfrac{2a}{t_1} + \dfrac{b}{t_2}} \tag{2.67}$$

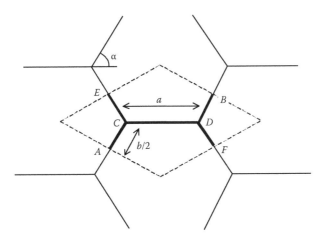

FIGURE 2.12
Hexagonal core.

For the computation of $\langle c_{2323} \rangle$, we can directly apply (2.58) to obtain piecewise constant shear flow along ACDB/ECDF; however, with the consideration of double symmetry the approach can be further simplified as follows. Since segment CD is geometrically neutral, segments ACE and BDF can be instead taken as periodic segments. Then, for either of these segments, d_2 and T can be expressed as

$$d_2 = b \sin \theta$$

$$T = \frac{b}{t_2} \qquad\qquad (2.68)$$

Then, using (2.62), we can simply write

$$\langle c_{2323} \rangle = \frac{G_L}{A} 2 t_2 b \sin^2 \theta \qquad\qquad (2.69)$$

2.2.4 Verification Using Finite Element Analysis

The formulas given in (2.67) and (2.69) for hexagonal honeycomb core configuration lead to identical results given by Kelsey et al. (1958), Gibson and Ashby (1988), and Shi and Tong (1995a). Thus, we only performed finite element (FE) analyses to confirm the results for the honeycomb core configurations with sinusoidal and tubular geometries.

The commercial finite element program ANSYS 5.5 is used, and eight-node isoparametric layered shell elements (SHELL 99) are selected to model the thin-walled cores. To minimize the computational effort, the FE model is developed with a special consideration of periodicity. Based on the periodic nature of honeycomb core structures, one-quarter of a unit cell (see Figures 2.13 and 2.14) is modeled with boundary conditions specified in Tables 2.13 and 2.14 for sinusoidal and tubular cores, respectively. The boundary conditions are specified as prescribed displacements v_i corresponding to coordinates y_i ($i = 1, 2, 3$). In the FE model of the sinusoidal core (Figure 2.13), a uniform displacement of v_1 is applied to all the nodes at the top surface defined by the curve line EO, and the resulting shear force F_1 along E′O′ can be correspondingly obtained. To keep a periodic boundary condition and obviate the bending effect, the displacement component along the vertical boundary EE′ and OO′ must remain linear from top to bottom by imposing $v_2 = 0$. In the numerical simulation, the parameters assumed for the sinusoidal core curve are listed in Table 2.15, and their physical meanings are given in the preceding section.

Similarly for the FE model of tubular core (Figure 2.14), either v_1 or v_2 is applied to all the nodes at the top surface (i.e., for the curve lines EO and EC), and the resulting shear force, F_1 or F_2, along the E′O′ and E′C′ is

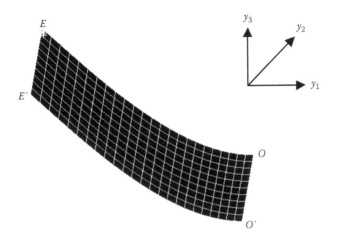

FIGURE 2.13
FE model of a quarter of unit cell for sinusoidal core.

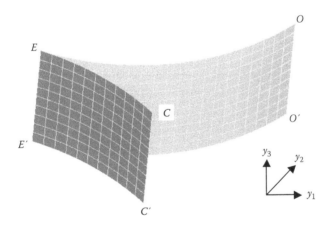

FIGURE 2.14
FE model of a quarter of unit cell for tubular core.

TABLE 2.13

Boundary Conditions in FE Mode of Quarter Unit Cell of Sinusoidal Core

Specified Displacement on Boundary	v_1	v_2	v_3
EE'/OO'	Linear	0	0
EO	0.0001	0	Free
E'O'	0	0	Free

Note: For the computation of $\langle c_{1313} \rangle$.

TABLE 2.14

Boundary Conditions in FE Model of a Quarter Unit Cell of Tubular Core

Specified Displacement on Boundary	v_1	v_2	v_3
EE′/OO′/CC′	Linear	0	0
EO/EC	0.0001	0	Free
E′O′/E′C′	0	0	Free

Note: For the computation of $\langle c_{1313} \rangle$.

TABLE 2.15

Comparison between Analytical and Numerical Predictions for Sinusoidal Core

	Parameters		$\langle c_{1313} \rangle$
Numerical	$v_1 = 0.0001$	$A = 2ab = 8;$	$\dfrac{4F_1}{A\gamma_1} = 3416$
	$h = 0.5$	$y_1 = \dfrac{u_1}{2h} = 0.0002;$	
	$a = b = 2$	$F_1 = 1.3664;$	
	$t = 0.01$		
Analytical		$S = 2.9274; G_L = 10^6$	$\dfrac{at_2}{bS} G_L = 3416$

Note: The computation is based on the curve portion shown in Figure 2.13.

correspondingly computed. Again, to prevent the bending effect, the displacement component at vertical boundaries EE′, OO′, and CC′ must remain linear from top to bottom by imposing $v_2 = 0$. The parameters used in the FE model for a tubular core are specified in Table 2.16.

The elastic properties of the core wall of the shell elements (SHELL 99) can be modeled as orthotropic, and a simple parametric study of elastic stiffness shows that the resulting shear force $F_i (i = 1, 2)$ is only dependent on the variable G_L. This is consistent with the derivation of (2.53) through (2.57).

By obtaining the resulting shear force from the FE models, the effective shear stiffness can be calculated, and comparisons are made with the analytical solution. The equations and calculations for sinusoidal and tubular cores are shown in Tables 2.15 and 2.16, respectively, which illustrate the accuracy of formula (2.62). From a structural mechanics point of view, the exactness of results is well expected once the appropriate boundary constrains are applied; i.e., the condition of constant shear flow is imposed so that identical solutions result from both the analytical and numerical approaches.

2.2.5 Summary and Discussions

With the application of (2.62), the formulas of transverse shear stiffness for arbitrary periodic cellular honeycomb cores can be easily obtained, and they

TABLE 2.16

Comparison between Analytical and Numerical Predictions for Tubular Core

	Parameters		$\langle c_{1313}\rangle$	$\langle c_{2323}\rangle$
Numerical	$A = 2\sqrt{3}R^2 = 2\sqrt{3}$;		$\dfrac{4F_1}{A\gamma_1} = 17332$	$\dfrac{4F_2}{A\gamma_2} = 16541$
	$v_1 = v_2 = 0.0001$;	$\gamma_1 = \gamma_2 = 0.0002$;		
	$h = 0.5$;	$F_1 = 3.0020$;		
	$R = 1$;	$F_2 = 2.8649$		
Analytical	$t = 0.02$;	$G_L = 10^6$	$\dfrac{(51 - 24\sqrt{3})t}{2\sqrt{3}\pi R}G_L = 17332$	$\dfrac{9t}{2\sqrt{3}\pi R}G_L = 16540$

would be valuable for design and optimization of honeycomb shape configurations (Qiao et al. 2008a,b). As a summary, the transverse shear stiffness formulas for several common core configurations are given in Table 2.17, and they are expressed in terms of apparent core density ρ_c.

It must be noted that the formulation based on the above homogenization process actually corresponds to the lower bound of transverse shear stiffness of sandwich structures, for which the facesheet warping constraints have not been taken into account. As indicated by other researches on hexagonal honeycombs (Grediac 1993; Shi and Tong 1995a), the facesheet effect is quite localized, especially for sandwich cores with a moderate ratio of unit cell size to core height. The upper bound of transverse shear stiffness can be obtained by the assumption of infinitely stiff facesheets and the principle of minimum energy theorem (e.g., for hexagonal cores by Kelsey et al. 1958; Gibson and Ashby 1988). The upper bound of transverse shear stiffness for general core configuration can be expressed as

$$\langle c_{1313}\rangle = \frac{G_L}{\Lambda}\sum_k \int_A^B \cos^2\theta\, t\, ds$$

$$\langle c_{2323}\rangle = \frac{G_L}{\Lambda}\sum_k \int_A^B \sin^2\theta\, t\, ds$$

(2.70)

As an illustration, the upper bounds of transverse shear stiffness are also given in Table 2.17 for common core configurations. Note that in Table 2.17 the exact solutions are attained for several core configurations (e.g., for triangle and rectangular grid cores) due to the identical lower and upper bounds of transverse shear stiffness, as one would intuitively expect.

To further precisely evaluate the facesheet effect or obtain a narrow bound solution range, the energy method can be used as an effective approach whenever the additional local energy is rationally weighed. Based on the

TABLE 2.17

Transverse Shear Stiffness of Common Honeycomb Core Configurations

Honeycomb		$\langle c_{1313}^{Lower}\rangle/G_L$ Equation (2.62)	$\langle c_{1313}^{Upper}\rangle/G_L$ Equation (2.70)	$\langle c_{2323}^{Lower}\rangle/G_L$ Equation (2.62)	$\langle c_{2323}^{Upper}\rangle/G_L$ Equation (2.70)	2D Shape Diagram (constant core wall thickness)
Regular hexagon ($a = b$)		$\dfrac{8}{9}\cos^4\dfrac{\theta}{2}\rho_c$	$\dfrac{2\cos^2\theta+1}{3}\rho_c$	$\dfrac{2}{3}\sin^2\theta\,\rho_c$	$\dfrac{2}{3}\sin^2\theta\,\rho_c$	
Regular re-entrant		$\dfrac{8}{9}\cos^4\dfrac{\theta}{2}\rho_c$	$\dfrac{2\cos^2\theta+1}{3}\rho_2$	$\dfrac{2}{3}\sin^2\theta\,\rho_c$	$\dfrac{2}{3}\sin^2\theta\,\rho_c$	
Reinforced sinusoidal	$a = 0.5b$	$0.4339\,\rho_c$	$0.4555\,\rho_c$	$0.5251\,\rho_c$	$0.5445\,\rho_c$	
	$a = b$	$0.6832\,\rho_c$	$0.6963\,\rho_c$	$0.2773\,\rho_c$	$0.3037\,\rho_c$	
	$a = 2b$	$0.8773\,\rho_c$	$0.8803\,\rho_c$	$0.1025\,\rho_c$	$0.1197\,\rho_c$	
Triangle grid		$\cos\theta\rho_c$	$\cos\theta\rho_c$	$2\sin^2\dfrac{\theta}{2}\rho_c$	$2\sin^2\dfrac{\theta}{2}\rho_c$	
Square packing tubular		$\dfrac{4}{\pi^2}\rho_c$	$\dfrac{1}{2}\rho_c$	$\dfrac{4}{\pi^2}\rho_c$	$\dfrac{1}{2}\rho_c$	
Hexagonal packing tubular		$\dfrac{51-24\sqrt{3}}{2\pi^2}\rho_c$	$\dfrac{1}{2}\rho_c$	$\dfrac{9}{2\pi^2}\rho_c$	$\dfrac{1}{2}\rho_c$	
Rectangular grid		$\dfrac{a}{a+b}\rho_c$	$\dfrac{a}{a+b}\rho_c$	$\dfrac{b}{a+b}\rho_c$	$\dfrac{b}{a+b}\rho_c$	

principle of minimum energy theorem, the consideration of complex interactions between facesheets and core can be simplified by assuming an appropriate displacement field, which can result in an improved upper bound solution estimation. By the Raleigh–Ritz method, there are basically two approaches, one to specify the facesheet displacement field, and the other the core internal displacement field. The latter, successfully implemented by Penzien and Didriksson (1964) for the hexagonal core, seems more effective and is suggested to apply for the present formulation for all general core configurations.

2.2.6 Conclusions

In this section, an analytical approach using a two-scale asymptotic homogenization technique is presented for effective transverse shear stiffness evaluations of honeycomb structures, and an explicit formula is provided for general shapes of thin-walled honeycomb cores. Three typical core configurations are subsequently solved with the developed formula, which is validated by existing or numerical solutions. The derived formula (2.62) can be efficiently used to predict the effective transverse shear stiffness of honeycomb cores with any general core configurations and can be applied to optimization of honeycomb core structures (Qiao et al. 2008a,b). Further, this approach, with certain modifications, can be extended to other sandwich structures, including the consideration of wall thickness effect for thick wall cores.

2.3 Homogenized Elastic Properties of Honeycomb Sandwiches with Skin Effect

In this section, the homogenized elastic properties of a hexagonal honeycomb core with consideration of skin effect (Xu and Qiao 2002; Quao and Xu 2005) are evaluated, and as in most existing studies of honeycomb core properties in sandwich construction, the skin effect is mostly not accounted for.

2.3.1 Introduction

The natural efficiency of cellular structures has thus attracted many investigations (e.g., Gibson and Ashby 1988; Warren and Kraynik 1987; Fortes and Ashby 1999) on periodic and disordered cells, wherein the book of Gibson and Ashby (1988) is the first systematic literature in the field. Of the fundamental equivalent elastic properties, the in-plane elastic properties of honeycomb were first obtained with the standard beam theory (Gibson and Ashby 1988; Masters and Evans 1996). Further refinements, e.g., as introduced by

Masters and Evans (1996), have been attempted considering stretching and hinging effects and the extension to a finer scale of molecular modeling. It must be pointed out that all these mathematical models on honeycomb cores are built based on pure cellular structures, and the usual presence of strengthening skin faces has not been taken into account. In classical sandwich theory (Allen 1969), the global skin-core interaction is identified as the result of the antiplane core assumption. Since the constraints of two skin faces significantly alter the local deformation mechanism of a hetero-geneous core, the homogenized core stiffness properties become sensitive to the ratio of core thickness to unit cell size, which is called skin effect in this chapter or thickness effect by Becker (1998). The ignorance of skin effect has been prevalent in today's sandwich research and design, wherein the equivalent core properties are simply taken from those formulas based on pure cellular models. Besides other unscrupulous uses causing erroneous Poisson's ratios and singularities, this ignorance yields an underestimate of stiffness and subsequent inconsistencies between modeling and real-ity, though only a few of them were noticed in experiments (e.g., the study by Cunningham and White (2001)). A common example is the antiplane core assumption in sandwich beam analysis, where skin effect and edge effect of anticlastic bending have been too simply ignored. As observed in experiments (e.g., Adams and Maheri 1993; Daniel and Abot 2000), skin con-straints were demonstrated by the phenomenon of skin lateral contraction and expansion.

The skin effect, induced by a high gradient of material change between two skin faces and a heterogeneous core, can be analogous to heterogeneous multiphase interactions in micromechanics of composites. The homogeniza-tion theory, well applied in 3D microscale periodic composite materials, has been adapted into the heterogeneous plate and shell theory since the 1970s (Duvaut 1977; Caillerie 1984). The theoretical efforts made in obtaining plate equivalent properties are highly dependent on the simplifications given by the constraint assumptions of a corresponding plate theory, i.e., Kirchhoff, Reissner–Hencky, or Reddy plate theories (Lewinski 1991). And the approxi-mations are processed based on the ratio of a plate's two small parameters, i.e., characteristic thickness δ and characteristic periodicity ε. When $\delta/\varepsilon \sim 0$, such as lattice plate, the plate assumption of a unit cell results in simple ana-lytical formulas (Lewinski 1991; Caillerie 1984). When $\delta/\varepsilon \gg 1$, as in the case of fiber composite laminates, the equivalent stiffness properties of laminates are derived from the micromechanics between fiber and a matrix in each indi-vidual sublayer, such as classical laminate plate theory (CLPT) resulting from the plane stress assumption. In the case of $\delta/\varepsilon \sim 1$, the asymptotic expansion method or G convergence technique generates the Caillerie–Kohn–Vogelius plate model, which is difficult to apply analytically (Lewinski 1991). Hence, honeycomb sandwiches (e.g., for the one shown in Figure 2.1), fallen into the domain of $\delta/\varepsilon \sim 1$, have been conventionally treated by laminate theory, where a honeycomb core is first homogenized into a continuum equivalent layer separately, and then the skin-core interactions are modeled with CLPT or higher-order laminate theories. Clearly, even with higher-order terms, the conventional approach fails to realize the heterogeneity of cores and the consequent high through-thickness mechanical variables.

In conventional sandwich analysis, the three-layer sandwich theory requires equivalent properties of a pure core, which should be accountable for real skin-core interactions in both a global and a local sense; i.e., the interactions must be energetically equivalent prior to and posterior to the homogenization of the core. There have been many refined theories and finite elements proposed to overcome the complicated interaction problem. In this section, a straightforward approach is proposed to homogenize a unit cell, including both skin faces and core, by which skin effect cannot only be accounted for locally, but also be assessed precisely. A homogenized single-layer plate model then can be constructed with the properties derived based on shear deformable plate theory. Further, with this approach the sandwich local behaviors can be accurately predicted by using an inverse or unsmearing procedure, which is expected as an important advantage over any conventional refined theories. One example is of the local stability problem, where the critical wrinkling load is strongly dependent on the in-plane stiffness (Vonach and Rammerstorfer 1998). The diagram of comparison between conventional and proposed approaches is shown in Figures 2.15 and 2.16.

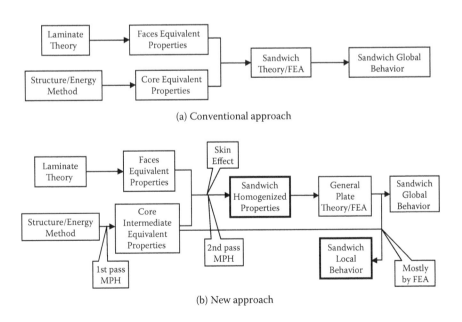

(a) Conventional approach

(b) New approach

FIGURE 2.15
Diagram of two approaches for honeycomb sandwich analysis.

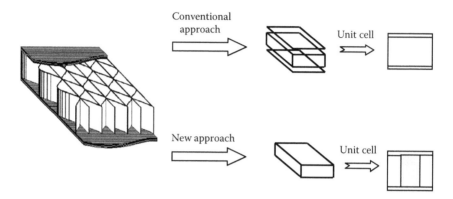

FIGURE 2.16
Comparison of two approaches for honeycomb sandwich analysis.

2.3.2 Literature Review

The phenomenon of honeycomb skin effect was first assessed by Kelsey et al. (1958), and the equivalent transverse shear stiffness of the hexagonal honeycomb core was investigated. The skin effect on the transverse shear deformation was theoretically expressed by two bounds of derived equivalent shear stiffness, whereas the lower and upper bounds correspond to zero and an infinitely large skin effect, respectively. Later, Penzien and Didriksson (1964) improved Kelsey's upper bound by formulating a closer displacement field than Kelsey's uniform field, and the solution consistently showed the trend of the diminishing skin effect with the increase of core thickness ratio δ/ε. More recently, Shi and Tong (1995a) applied a 2D homogenization technique to obtain Kelsey's lower bound value, and Xu et al. (2001) extended it to general honeycomb configurations. Numerical approaches on transverse shear stiffness were attempted by Shi and Tong (1995a) and Grediac (1993). Shi and Tong (1995a) used a specialized hybrid element; however, the results found are consistent with neither Penzien and Didriksson's (1964) conclusion nor Saint-Venant's theorem explained by Grediac (1993). In the study by Grediac (1993), the applied unit cell boundary conditions actually correspond to those of the analytical approximations of Penzien and Didriksson (1964); thus, the results only numerically verify the latter's work without giving more accuracies.

Unlike the evaluations of equivalent shear stiffness of honeycomb cores (Qiao and Xu 2005), the skin effect on other stiffness tensors has received few attentions. Both Parton and Kudryavtsev (1993) and Shi and Tong (1995b) solved the honeycomb equivalent in-plane stiffness by a two-scale method; however, the skin effect was not considered in their study. Through an energy minimization implementation, Becker (1998), for the first time, assessed the skin effect on equivalent in-plane moduli. Further expansion was attempted recently by Hohe and Becker (2001), to include all elastic tensors and general honeycomb cores, but with lengthy and implicit calculations. For a sandwich panel, the very important concern is the flexural stiffness contributed from a honeycomb core. Among few existing literatures, only Parton and Kudryavtsev (1993) gave a formula for flexural stiffness of honeycomb cores, which is simply derived from in-plane stretch stiffness and does not include skin effect. In this section, the flexural stiffness is distinguished from stretch stiffness, since the displacement field varies when loading is changed from symmetry to antisymmetry about the panel middle plane.

There are several engineering investigations (Bourgeois et al. 1998; Chamis et al. 1988; Vougiouka and Guedes 1998; Takano et al. 1995) on the homogenization of honeycomb sandwiches; however, they are case limited without further insight. Besides addressing the aforementioned unsolved issues, this section aims to introduce an effective approach to homogenize general honeycomb cells and to provide a comprehensive approach in three aspects: the mathematical statement of the sandwich homogenization theory, the analytical solution of a multipass homogenization (MPH) technique, and a 3D unit cell Finite Element Analysis (FEA) homogenization technique.

The MPH technique originates from the meditation that the homogenization of an object can be processed by its principal axes one by one; i.e., the homogenized results obtained along one axis can be well applied to the next pass along another axis. In this section, the MPH technique includes a two-pass procedure to homogenize 3D geometrical heterogeneous honeycomb media in orthogonal directions. The first pass involves the building of a geometry-to-material transformation model (GTM), by which the complicated 3D spatial analysis is simplified into a 2D plane stress or plane strain case. The first pass, i.e., the GTM model, is mathematically equivalent to the coordinate transformation in combination with energetic averaging. The coordinate transformation has been conventionally applied by all the relevant research referenced in this section, but the process is inconvenient and lengthy. The MPH technique and sandwich homogenization formulation introduced here can efficiently simplify the process and be applicable for all general sandwich structures with periodic cores. In the second pass, with the resulting intermediate core equivalent properties, the appropriate displacement field is constructed by satisfying field equations either exactly or weakly. With the homogenization formulation given in Section 2.3.3, or with the energy minimization theorem, the homogenized stiffness can be analytically solved in the form of Fourier series. In Section 2.3.5, the FEA numerical results verify the semianalytical solutions, which further complementally show the influence of skin rigidity. A specialized FE modeling technique is introduced at the end of this section by appropriately imposing periodic boundary conditions in unit cell FEA modeling. This technique can be easily used in commercial FEA packages without writing specialized code for hybrid elements (Manet et al. 1998), the principle of which is extendable to all periodic media of unit cell FEA.

2.3.3 Formulation of Honeycomb Homogenization Problem

2.3.3.1 Asymptotic Expansions about Plate Thickness δ

The asymptotic expansion for plates with $\delta/\varepsilon \sim 1$ was first given by Caillerie (1984). Hereby the expansion is repeated, and the notations are made consistent with the derivation of the Kirchoff-Love plate model by Parton and Kudryavtsev (1993), where small Latin indices denote 1, 2, 3 and small Greek indices for 1, 2. To extend the homogenized plate model to transverse shear deformation theory, the formulation of homogenized transverse shear stiffness is attempted in Section 2.3.3.2.

For common honeycomb sandwiches (see Figure 2.8) of which the body force is ignored, the 3D elasticity field equations and boundary conditions can be written as

$$\frac{\partial \sigma_{ij}}{\partial x_j} = 0, \quad \text{in } \Omega$$

$$u_i = \hat{u}_i, \quad \text{on } \partial_1 \Omega \tag{2.71}$$

$$\sigma_{ij} n_j = T_i, \quad \text{on } \partial_2 \Omega$$

where

$$\sigma_{ij} = c_{ijkl} e_{kl}$$

$$e_{kl} = \frac{1}{2} \left(\frac{\partial u_k}{\partial x_l} + \frac{\partial u_l}{\partial x_k} \right) \tag{2.72}$$

And the coefficient c_{ijkl} should satisfy the elliptical symmetry condition

$$c_{ijkl} = c_{jikl} = c_{ijlk} = c_{klij}$$

The transverse thickness of the sandwiches is denoted δ, and hereby the local coordinates of the unit cell are introduced to rescale the problem:

$$y_1 = \frac{x_1}{h_1 \delta} \qquad y_2 = \frac{x_2}{h_2 \delta} \qquad z = \frac{x_3}{\delta} \tag{2.73}$$

Thus, the rescaled unit cell domain is in $\{-1/2 < y_1, y_2, z < 1/2\}$. The external traction T is assumed to only be applied in the transverse direction, which can be written in the function of δ by

$$T_3^{\pm z}(x) = \delta^3 q^{\pm z}(x, y) \tag{2.74}$$

By matching the expansion terms, the boundary traction conditions are known:

$$\sigma_{i3}^{(l)}\big|_{z=\pm\frac{1}{2}} = 0, \quad l=0,1,2$$

$$\sigma_{13}^{(3)}\big|_{z=\pm\frac{1}{2}} = \sigma_{23}^{(3)}\big|_{z=\pm\frac{1}{2}} = 0, \quad \sigma_{33}^{(3)}\big|_{z=\pm\frac{1}{2}} = q^{\pm z}(x,y)$$

(2.75)

With the two-scale expansion method about the small parameter δ, the series are expressed as

$$u_i^{(\varepsilon)} = u_i^{(0)}(x) + \delta\, u_i^{(1)}(x,y) + \cdots$$

$$e_{ij}^{(\varepsilon)} = e_{ij}^{(0)}(x,y) + \delta\, e_{ij}^{(1)}(x,y) + \cdots$$

(2.76)

$$\sigma_{ij}^{(\varepsilon)} = \sigma_{ij}^{(0)}(x,y) + \delta\, \sigma_{ij}^{(1)}(x,y) + \cdots$$

where the strain-displacement law is

$$e_{\alpha\beta}^{(l)} = \varepsilon_{\alpha\beta}^{(l)} + \frac{1}{2}\left(\frac{1}{h_\beta}\frac{\partial u_\alpha^{(l+1)}}{\partial y_\beta} + \frac{1}{h_\alpha}\frac{\partial u_\beta^{(l+1)}}{\partial y_\alpha}\right)$$

$$e_{\alpha 3}^{(l)} = \frac{1}{2}\varepsilon_{\alpha 3}^{(l)} + \frac{1}{2}\left(\frac{\partial u_\alpha^{(l+1)}}{\partial z} + \frac{1}{h_\alpha}\frac{\partial u_3^{(l+1)}}{\partial y_\alpha}\right)$$

$$e_{33}^{(l)} = \frac{\partial u_3^{(l+1)}}{\partial z}$$

(2.77)

$$\varepsilon_{\alpha\beta}^{(l)} = \frac{1}{2}\left(\frac{\partial u_\alpha^{(l)}}{\partial x_\beta} + \frac{\partial u_\beta^{(l)}}{\partial x_\alpha}\right)$$

$$\varepsilon_{\alpha 3}^{(l)} = \frac{\partial u_3^{(l)}}{\partial x_\alpha}$$

$$l = 0,1,2,\cdots$$

and the constitutive equations are

$$\sigma_{ij}^{(l)} = c_{ijkl}e_{kl}^{(l)}$$

(2.78)

$$l = 0,1,2,\cdots$$

Note the variables with superscript $l > 0$ are all Y-periodic in y. Substituting (2.76), (2.77), and (2.78) into (2.71) and matching the power order of δ results

$$\frac{1}{h_\alpha}\frac{\partial\sigma_{i\alpha}^{(0)}}{\partial y_\alpha}+\frac{\partial\sigma_{i3}^{(0)}}{\partial z}=0$$

$$\frac{\partial\sigma_{i\alpha}^{(l)}}{\partial x_\alpha}+\frac{1}{h_\alpha}\frac{\partial\sigma_{i\alpha}^{(l+1)}}{\partial y_\alpha}+\frac{\partial\sigma_{i3}^{(l+1)}}{\partial z}=0,\, l=0,1,2,\cdots$$

(2.79)

With the substitution of (2.76), (2.77), and (2.78) into (2.79) and the consideration of the boundary conditions in (2.75), the following can be derived as detailed by Parton and Kudryavtsev (1993):

$$u_\alpha^{(0)}=0$$

$$u_3^{(0)}(x)=w(x)$$

$$u_\alpha^{(1)}=v_\alpha^{(1)}-z\frac{\partial w}{\partial x_\alpha}$$

$$\sigma_{ij}^{(0)}=0$$

(2.80)

from which the displacements can be expressed by global in-plane and flexural variables as

$$u_\alpha^{(\varepsilon)}=\delta v_\alpha^{(1)}(x)-x_3\frac{\partial w}{\partial x_\alpha}+\delta^2 U_\alpha^{\mu\nu}\varepsilon_{\mu\nu}^{(1)}+\delta^2 V_\alpha^{\mu\nu}\kappa_{\mu\nu}^{(0)}+\cdots$$

$$u_3^{(\varepsilon)}=w(x)+\delta^2\tilde{u}_3^{(2)}(x)+\delta^2 U_3^{\mu\nu}\varepsilon_{\mu\nu}^{(1)}+\delta^2 V_3^{\mu\nu}\kappa_{\mu\nu}^{(0)}+\cdots$$

(2.81)

where

$$\kappa_{\mu\nu}^{(0)}=-\frac{\partial^2 w}{\partial x_\mu\,\partial x_\nu}$$

$$\varepsilon_{\mu\nu}^{(1)}=\frac{1}{2}\left(\frac{\partial v_\mu^{(1)}}{\partial x_\nu}+\frac{\partial v_\nu^{(1)}}{\partial x_\mu}\right)$$

(2.82)

Note in Parton and Kudryavtsev (1993) that the transverse shear deformation is ignored, while in this section the additional term $\delta^2\tilde{u}_3^{(2)}(x)$ in (2.81) is obtained by the modification of Equation (19.51) of Parton and Kudryavtsev (1993) as follows:

$$u_3^{(2)}(x,y,z)=U_3^{\mu\nu}\varepsilon_{\mu\nu}^{(1)}+V_3^{\mu\nu}\kappa_{\mu\nu}^{(0)}+\tilde{u}_3^{(2)}(x)$$

(2.83)

In the above equations, the commonly called homogenization functions $U(y,z)$ and $V(y,z)$ are the local periodic displacements induced by in-plane strain

$\varepsilon_{\mu\nu}^{(1)}$ and flexural curvature $\kappa_{\mu\nu}^{(0)}$, respectively, which should satisfy the local equilibrium equations:

$$\left\langle \sigma_{\mu\nu}^{(1)} \right\rangle = \left\langle C_{\mu\nu\alpha\beta} \right\rangle \varepsilon_{\alpha\beta}^{(1)} + \left\langle C_{\mu\nu\alpha\beta}^* \right\rangle \kappa_{\alpha\beta}^{(0)}$$

$$\left\langle z\sigma_{\mu\nu}^{(1)} \right\rangle = \left\langle zC_{\mu\nu\alpha\beta} \right\rangle \varepsilon_{\alpha\beta}^{(1)} + \left\langle zC_{\mu\nu\alpha\beta}^* \right\rangle \kappa_{\alpha\beta}^{(0)}$$

(2.84)

$$C_{\mu\nu\alpha\beta} = c_{\mu\nu\alpha\beta} + c_{\mu\nu i\lambda} \frac{1}{h_\lambda} \frac{\partial U_i^{\alpha\beta}}{\partial y_\lambda} + c_{\mu\nu i3} \frac{\partial U_i^{\alpha\beta}}{\partial z}$$

$$C_{\mu\nu\alpha\beta}^* = z c_{\mu\nu\alpha\beta} + c_{\mu\nu i\lambda} \frac{1}{h_\lambda} \frac{\partial V_i^{\alpha\beta}}{\partial y_\lambda} + c_{\mu\nu i3} \frac{\partial V_i^{\alpha\beta}}{\partial z}$$

(2.85)

and uniqueness conditions

$$\left\langle U_i^{\alpha\beta} \right\rangle_y = 0, \quad z = 0$$

$$\left\langle V_i^{\alpha\beta} \right\rangle_y = 0, \quad z = 0$$

(2.86)

where the averaging operator $\langle \times \rangle = \int_{\Omega_p} \times \, dy_1 dy_2 dz$, $\langle \times \rangle_y = \int_Y \times \, dy_1 dy_2$.

2.3.3.2 Homogenized Transverse Shear Plate Model

To take into account the transverse shear deformation, instead of (2.86), the uniqueness conditions are redefined as

$$\left\langle c_{\alpha 3\beta 3} U_3^{\alpha\beta} \right\rangle = 0$$

$$\left\langle c_{\alpha 3\beta 3} V_3^{\alpha\beta} \right\rangle = 0$$

(2.87)

By applying the averaging process to (2.79) for $l = 2$, it gives

$$\frac{\partial \left\langle \sigma_{3\alpha}^{(2)} \right\rangle}{\partial x_\alpha} + q(x) = 0$$

(2.88)

$$q(x) = \int_Y q^{\pm z}(x, y) dy$$

(2.89)

Assume material properties are monoclinic, that is, $c_{\alpha\beta\mu 3} = c_{\alpha 333} = 0$. Using (2.77) and (2.78), we have

$$\sigma_{3\alpha}^{(2)} = c_{3\alpha3\beta}(\frac{\partial u_3^{(2)}(x,y,z)}{\partial x_\beta} + \frac{\partial u_\beta^{(3)}}{\partial z} + \frac{1}{h_\beta}\frac{\partial u_3^{(3)}}{\partial y_\beta}) \tag{2.90}$$

Let

$$\gamma_{3\beta}^{(2)} = \frac{\partial \tilde{u}_3^{(2)}(x)}{\partial x_\beta} \tag{2.91}$$

$$u_i^{(3)} = \Pi_i^\beta \gamma_{3\beta}^{(2)}$$

and define equivalent transverse shear stiffness $<C_{\alpha3\beta3}>$ by

$$\left\langle \sigma_{3\alpha}^{(2)} \right\rangle = \left\langle C_{\alpha3\beta3} \right\rangle \gamma_{3\beta}^{(2)} \tag{2.92}$$

Then, by (2.83), (2.87), and (2.90) through (2.92) and even about z, $C_{\alpha3\beta3}$ can be written as

$$C_{\alpha3\beta3} = c_{\alpha3\beta3} + c_{\alpha3\lambda3}(\frac{\partial \Pi_\lambda^\beta}{\partial z} + \frac{1}{h_\lambda}\frac{\partial \Pi_3^\beta}{\partial y_\lambda}) \tag{2.93}$$

Thus, (2.85) and (2.92) give all the stiffness required for transverse shear deformation theory of plates. And the stress and moment resultants can be written as

$$N_{\alpha\beta} = \delta^2 \left\langle C_{\alpha\beta\mu\nu} \right\rangle \varepsilon_{\mu\nu}^{(1)} + \delta^2 \left\langle C_{\alpha\beta\mu\nu}^* \right\rangle \kappa_{\mu\nu}^{(0)}$$

$$M_{\alpha\beta} = \delta^3 \left\langle z C_{\alpha\beta\mu\nu} \right\rangle \varepsilon_{\mu\nu}^{(1)} + \delta^3 \left\langle z C_{\alpha\beta\mu\nu}^* \right\rangle \kappa_{\mu\nu}^{(0)} \tag{2.94}$$

$$Q_\alpha = \delta^3 \left\langle C_{\alpha3\beta3} \right\rangle \gamma_{3\beta}^{(2)}$$

For transversely symmetric honeycomb sandwiches, clearly the in-plane and flexural couple disappears, i.e., $\left\langle z C_{\alpha\beta\mu\nu} \right\rangle = \left\langle C_{\alpha\beta\mu\nu}^* \right\rangle = 0$. From (2.81), the global strain variables are correspondingly given by

$$\gamma_{3\beta} = \delta^2 \gamma_{3\beta}^{(2)}$$

$$\varepsilon_{\mu\nu} = \delta \varepsilon_{\mu\nu}^{(1)} \tag{2.95}$$

$$\kappa_{\mu\nu} = \kappa_{\mu\nu}^{(0)}$$

Further for symbolic consistence with the classical laminate plate theory and by (2.94) and (2.95), the plate macroscopic equivalent stiffness may be finally expressed as

$$N_{\alpha\beta} = A^H_{\alpha\beta\mu\nu}\varepsilon_{\mu\nu}, \qquad A^H_{\alpha\beta\mu\nu} = \delta \left\langle C_{\alpha\beta\mu\nu} \right\rangle$$

$$M_{\alpha\beta} = D^H_{\alpha\beta\mu\nu}\kappa_{\mu\nu}, \qquad D^H_{\alpha\beta\mu\nu} = \delta^3 \left\langle zC^*_{\alpha\beta\mu\nu} \right\rangle \qquad (2.96)$$

$$Q_{\alpha} = H^H_{\alpha3\beta3}\gamma_{3\beta}, \qquad H^H_{\alpha3\beta3} = \delta \left\langle C_{\alpha3\beta3} \right\rangle$$

or more clearly for sandwich panels with three planes of symmetry:

$$\begin{bmatrix} N_{11} \\ N_{22} \\ N_{66} \end{bmatrix} = \begin{bmatrix} A_{11} & A_{12} & 0 \\ A_{21} & A_{22} & 0 \\ 0 & 0 & A_{66} \end{bmatrix} \begin{bmatrix} \varepsilon_{11} \\ \varepsilon_{22} \\ \varepsilon_{66} \end{bmatrix}$$

$$\begin{bmatrix} M_{11} \\ M_{22} \\ M_{66} \end{bmatrix} = \begin{bmatrix} D_{11} & D_{12} & 0 \\ D_{21} & D_{22} & 0 \\ 0 & 0 & D_{66} \end{bmatrix} \begin{bmatrix} \kappa_{11} \\ \kappa_{22} \\ \kappa_{66} \end{bmatrix} \qquad (2.97)$$

$$\begin{bmatrix} Q_{44} \\ Q_{55} \end{bmatrix} = \begin{bmatrix} H_{44} & 0 \\ 0 & H_{55} \end{bmatrix} \begin{bmatrix} \gamma_{44} \\ \gamma_{55} \end{bmatrix}$$

where contracted notation is introduced with $\alpha\alpha\beta\beta = \alpha\beta$, $2323 = 3232 = 44$, $1313 = 3131 = 55$, and $1212 = 2121 = 66$.

2.3.3.3 Field Equations of Three Local Problems

From (2.85) and (2.93), it is obvious that the homogenized stiffness can be obtained once the solutions of the homogenization functions $U(y, z)$, $V(y, z)$, and $P(y, z)$ are known. These periodic functions then have to be solved by the local elastic equilibrium equations as given in (2.79), combined with unit cell periodic boundary conditions. It should be noted that in all the following equations, the material properties (c) are a function of spatial coordinates, and assumed monoclinic that $c_{\alpha\beta\mu3} = c_{\alpha333} = 0$.

2.3.3.3.1 Transverse Shear Local Problem P(y, z)

Assume a pure shear case where $e = k = 0$. Combining (2.79), (2.90), and (2.91), the equilibrium equations become

$$\frac{1}{h_\alpha}\frac{\partial}{\partial y_\alpha}\left[c_{\alpha3\beta3} + c_{\alpha3\mu3}(\frac{\partial \Pi^\beta_\mu}{\partial z} + \frac{1}{h_\mu}\frac{\partial \Pi^\beta_3}{\partial y_\mu}) \right] + \frac{\partial}{\partial z}(c_{33\nu\lambda}\frac{1}{h_\lambda}\frac{\partial \Pi^\beta_\nu}{\partial y_\lambda} + c_{3333}\frac{\partial \Pi^\beta_3}{\partial z}) = 0$$

$$\frac{1}{h_\mu}\frac{\partial}{\partial y_\mu}(c_{\alpha\mu33}\frac{\partial \Pi^\beta_3}{\partial z} + c_{\alpha\mu\nu\nu}\frac{1}{h_\nu}\frac{\partial \Pi^\beta_\nu}{\partial y_\nu}) + \frac{\partial}{\partial z}\left[c_{\alpha3\beta3} + c_{\alpha3\lambda3}(\frac{\partial \Pi^\beta_\lambda}{\partial z} + \frac{1}{h_\lambda}\frac{\partial \Pi^\beta_3}{\partial y_\lambda}) \right] = 0 \qquad (2.98)$$

From (2.75), the boundary conditions are specified as

$$c_{33\upsilon\lambda}\frac{1}{h_\lambda}\frac{\partial\Pi_\upsilon^\beta}{\partial y_\lambda}+c_{3333}\frac{\partial\Pi_3^\beta}{\partial z}=0 \qquad z=z^\pm$$

$$c_{\alpha3\beta3}+c_{\alpha3\lambda3}(\frac{\partial\Pi_\lambda^\beta}{\partial z}+\frac{1}{h_\lambda}\frac{\partial\Pi_3^\beta}{\partial y_\lambda})=0 \qquad z=z^\pm$$

(2.99)

2.3.3.3.2 In-Plane (Stretch and Shear) Local Problem U(y, z)

Assume a pure in-plane case that $\gamma=\kappa=0$. Equations (2.79) and (2.85) lead to

$$\frac{1}{h_\mu}\frac{\partial}{\partial y_\mu}\left[c_{\mu3\upsilon3}(\frac{\partial U_\mu^{\alpha\beta}}{\partial z}+\frac{1}{h_\upsilon}\frac{\partial U_3^{\alpha\beta}}{\partial y_\upsilon})\right]+\frac{\partial}{\partial z}(c_{33\alpha\beta}+c_{33\upsilon\lambda}\frac{1}{h_\lambda}\frac{\partial U_\upsilon^{\alpha\beta}}{\partial y_\lambda}+c_{3333}\frac{\partial U_3^{\alpha\beta}}{\partial z})=0$$

$$\frac{1}{h_\mu}\frac{\partial}{\partial y_\mu}(c_{\nu\mu\alpha\beta}+c_{\nu\mu33}\frac{\partial U_3^{\alpha\beta}}{\partial z}+c_{\nu\mu\upsilon\lambda}\frac{1}{h_\lambda}\frac{\partial U_\upsilon^{\alpha\beta}}{\partial y_\lambda})+\frac{\partial}{\partial z}\left[c_{\upsilon3\omega3}(\frac{\partial U_\omega^{\alpha\beta}}{\partial z}+\frac{1}{h_\omega}\frac{\partial U_3^{\alpha\beta}}{\partial y_\omega})\right]=0$$

(2.100)

The boundary conditions are

$$c_{33\alpha\beta}+c_{33\upsilon\lambda}\frac{1}{h_\lambda}\frac{\partial U_\upsilon^{\alpha\beta}}{\partial y_\lambda}+c_{3333}\frac{\partial U_3^{\alpha\beta}}{\partial z}=0 \qquad z=z^\pm$$

$$c_{\upsilon3\omega3}(\frac{\partial U_\omega^{\alpha\beta}}{\partial z}+\frac{1}{h_\omega}\frac{\partial U_3^{\alpha\beta}}{\partial y_\omega})=0 \qquad z=z^\pm$$

(2.101)

2.3.3.3.3 Flexural Local Problem V(y, z)

Similarly, the equilibrium equations and boundary conditions are given as

$$\frac{1}{h_\mu}\frac{\partial}{\partial y_\mu}\left[c_{\mu3\upsilon3}(\frac{\partial V_\mu^{\alpha\beta}}{\partial z}+\frac{1}{h_\upsilon}\frac{\partial V_3^{\alpha\beta}}{\partial y_\upsilon})\right]+\frac{\partial}{\partial z}(zc_{33\alpha\beta}+c_{33\upsilon\lambda}\frac{1}{h_\lambda}\frac{\partial V_\upsilon^{\alpha\beta}}{\partial y_\lambda}+c_{3333}\frac{\partial V_3^{\alpha\beta}}{\partial z})=0$$

$$\frac{1}{h_\mu}\frac{\partial}{\partial y_\mu}(zc_{\nu\mu\alpha\beta}+c_{\nu\mu33}\frac{\partial V_3^{\alpha\beta}}{\partial z}+c_{\nu\mu\upsilon\lambda}\frac{1}{h_\lambda}\frac{\partial V_\upsilon^{\alpha\beta}}{\partial y_\lambda})+\frac{\partial}{\partial z}\left[c_{\upsilon3\omega3}(\frac{\partial V_\omega^{\alpha\beta}}{\partial z}+\frac{1}{h_\omega}\frac{\partial V_3^{\alpha\beta}}{\partial y_\omega})\right]=0$$

(2.102)

$$zc_{33\alpha\beta}+c_{33\upsilon\lambda}\frac{1}{h_\lambda}\frac{\partial V_\upsilon^{\alpha\beta}}{\partial y_\lambda}+c_{3333}\frac{\partial V_3^{\alpha\beta}}{\partial z}=0 \qquad z=z^\pm$$

$$c_{\upsilon3\omega3}(\frac{\partial V_\omega^{\alpha\beta}}{\partial z}+\frac{1}{h_\omega}\frac{\partial V_3^{\alpha\beta}}{\partial y_\omega})=0 \qquad z=z^\pm$$

(2.103)

2.3.4 Analytical Approach—Multipass Homogenization (MPH) Technique

In engineering applications of the homogenization theory, the exact analytical solutions are seldom obtainable and the approximations are usually made either by a semianalytical method or by pure numerical techniques, such as the finite element method. In the problem of honeycomb cells, there are 3D local functions (see Equations (2.98) through (2.103)) physically interpreted as a complicated combination of local warping, stretching, bending, shearing, twisting, etc. The exact mathematical expression for each of the functions is almost impossible to derive analytically. As demonstrated in (2.94), weak solutions are sufficient when the homogenized properties are sought in a variational sense.

For even a weak solution of the local problem as (2.98) through (2.103), their 3D deformations are difficult to deal with, and the direct construction of cell plates' displacement (Hohe and Becker 2001) involves a relatively complicated and implicit numerical process. Hereby in the first pass of the MPH, a simplified geometry-to-material transformation model (GTM) is proposed that a spatial heterogeneous problem can be transferred into a material heterogeneous problem with consequent intermediate equivalent properties. By this way, the strain energy of cell walls can be completely expressed by the resulting intermediate equivalent stiffness without omitting small higher-order terms, as exemplified by Φ_1 and Φ_2 in Equations (2.135) and (2.140) of Section 2.3.4.2.3. Note in the GTM model there is no restriction about the thickness of cell walls, as long as energy equivalence is satisfied with appropriate derivations of intermediate equivalent stiffness. In the second pass, the 2D heterogeneous problem then can be analytically homogenized in a unit cell by the variational approximations of the displacement field with the Rayleigh–Ritz method or partition method, etc. The weak form solution of the partial differential equations is finally verified with the FEA results in Section 2.3.5.

The MPH technique originates from the idea that the homogenization of an object may be processed by its principal axes one by one; i.e., the homogenized results obtained along one axis can be well applied to the second pass along another axis. There are several engineering applications of the MPH technique (e.g., Astley et al. 1997). The separation of the process is found very effective in the homogenization of honeycomb cells, as evidenced in this section. To illustrate the whole process explicitly, a better way is to follow an analytical example as given in Section 2.3.4, rather than a general procedure description. Hereby the most used hexagonal honeycomb is taken as an example (Figures 2.12 and 2.17), which has been given the most interest in the honeycomb family, with much available theoretical and experimental data.

2.3.4.1 First Pass—GTM Model

In principle of the GTM model, a doubly periodic sandwich panel is transformed into a one-dimensional periodic panel. For hexagons shown in Figure 2.17, the plate thus consists of two alternative thin sandwich beams with their intermediate equivalent material properties. In the first pass the GTM homogenization is made along, say x_2, so that the information of spatial periodicity along x_2 is stored in the consequent intermediate equivalent stiffness. The resulting 2D heterogeneous composite then can be conveniently assessed in the second pass along x_1.

For simplicity, the cell walls and skin faces are assumed both made of isotropic materials, where Young's modulus, shear modulus, and Poisson's ratio are denoted E_c, G_c, n_c and E_f, G_f, n_f, respectively. Geometrical notations of the hexagon are illustrated in Figure 2.12. It should be pointed out that for anisotropic materials the procedure is kept identical. The example is confined in transversely symmetric sandwiches with a thin-walled hexagonal core to which the thin beam model (Gibson and Ashby 1988) can be applied.

With the above assumptions, the faces of the two thin sandwich beams are known to be homogeneous, so that structural homogenization only needs to be processed on two types of beam cores, I and II (Figure 2.17). The intermediate equivalent properties may be obtained following the principle of structural mechanics, and the derivation refers to similar problems detailed

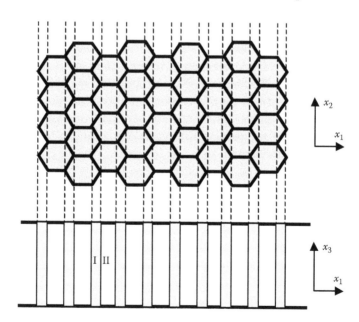

FIGURE 2.17
GTM—first pass of MPH technique.

in Gibson and Ashby (1988) and Masters and Evans (1996). The properties of core II consisting of parallel cell walls are

$$E_1^{hc2} = \frac{1}{2\sin\theta}\frac{t_2}{b}E_c, \quad E_2^{hc2} \approx 0, \quad E_3^{hc2} = \frac{1}{2\sin\theta}\frac{t_2}{b}E_c$$

$$G_{13}^{hc2} = \frac{1}{2\sin\theta}\frac{t_2}{b}G_c, \quad G_{23}^{hc2} \approx 0, \quad G_{12}^{hc2} = \frac{a/b+\cos\theta}{(a/b)^2(1+2a/b)\sin\theta}\left(\frac{t_2}{b}\right)^3 E_c \quad (2.104)$$

$$\nu_{12}^{hc2} \approx \nu_{21}^{hc2} \approx 0, \quad \nu_{31}^{hc2} = \nu_{13}^{hc2} = \nu_c, \quad \nu_{23}^{hc2} \approx \nu_{23}^{hc2} \approx 0$$

And the properties of core I with folded cell walls are

$$E_1^{hc1} = \frac{\cos\theta}{\sin^3\theta[1+ctg^2\theta(t_1/b)^2]}\left(\frac{t_1}{b}\right)^3 E_c$$

$$E_2^{hc1} = \frac{\sin\theta}{\cos^3\theta[1+tg^2\theta(t_1/b)^2]}\left(\frac{t_1}{b}\right)^3 E_c, \quad E_3^{hc1} = \frac{1}{\sin\theta\cos\theta}\frac{t_1}{b}E_c$$

$$G_{13}^{hc1} = ctg\theta\,\frac{t_1}{b}G_c, \quad G_{23}^{hc1} = tg\theta\,\frac{t_1}{b}G_c, \quad G_{12}^{hc1} = \sin\theta\cos\theta\,\frac{t_1}{b}E_c$$

$$\nu_{12}^{hc1} = ctg^2\theta\frac{1-(t_1/b)^2}{1+ctg^2\theta(t_1/b)^2}, \quad \nu_{21}^{hc1} = tg^2\theta\frac{1-(t_1/b)^2}{1+tg^2\theta(t_1/b)^2}$$

(2.105)

$$\nu_{31}^{hc1} = \nu_c\cos\theta, \quad \nu_{13}^{hc1} = \frac{\cos^3\theta}{\sin^2\theta[1+ctg^2\theta(t_1/b)^2]}\left(\frac{t_1}{b}\right)^2\nu_c$$

$$\nu_{32}^{hc1} = \nu_c\sin\theta, \quad \nu_{23}^{hc1} = \frac{\sin^3\theta}{\cos^2\theta[1+tg^2\theta(t_1/b)^2]}\left(\frac{t_1}{b}\right)^2\nu_c$$

where the superscript h denotes intermediate core equivalent properties, and c, 1, and 2 are for the core and its types I and II, respectively.

Note that unscrupulous calculation of in-plane Poisson's ratios, with such equations as (4.13) and (4.14) in Gibson and Ashby (1988), would result in in-plane stiffness much deviated from true values, and even produce singularities. The correct way is to further consider the stretch deformation, as noticed by Warren and Kraynik (1987) and Masters and Evans (1996). The formulas of Poisson's ratios in (2.105) are derivable from (4.50) in Gibson and Ashby (1988).

The generalized Hooke's law for the above hexagons with three planes of the elastic symmetry can be written in terms of the intermediate properties of principal elastic constants (Lekhnitskii 1968):

$$\varepsilon_1^h = \frac{1}{E_1^h}\sigma_1^h - \frac{v_{21}^h}{E_2^h}\sigma_2^h - \frac{v_{31}^h}{E_3^h}\sigma_3^h, \quad \gamma_{23}^h = \frac{1}{G_{23}^h}\tau_{23}^h$$

$$\varepsilon_2^h = -\frac{v_{12}^h}{E_1^h}\sigma_1^h + \frac{1}{E_2^h}\sigma_2^h - \frac{v_{32}^h}{E_3^h}\sigma_3^h, \quad \gamma_{13}^h = \frac{1}{G_{13}^h}\tau_{13}^h \tag{2.106}$$

$$\varepsilon_3^h = -\frac{v_{13}^h}{E_1^h}\sigma_1^h - \frac{v_{23}^h}{E_2^h}\sigma_2^h + \frac{1}{E_3^h}\sigma_3^h, \quad \gamma_{12}^h = \frac{1}{G_{12}^h}\tau_{12}^h$$

And the inverse of (2.106) results in the expression of stress variables:

$$\sigma_1^h = \frac{E_1^h}{1-\Delta}\left[(1-v_{23}^h v_{32}^h)\varepsilon_1^h + (v_{21}^h + v_{23}^h v_{31}^h)\varepsilon_2^h + (v_{31}^h + v_{21}^h v_{32}^h)\varepsilon_3^h\right], \quad \tau_{23}^h = G_{23}^h \gamma_{23}^h$$

$$\sigma_2^h = \frac{E_2^h}{1-\Delta}\left[(1-v_{13}^h v_{31}^h)\varepsilon_2^h + (v_{12}^h + v_{13}^h v_{32}^h)\varepsilon_1^h + (v_{32}^h + v_{12}^h v_{31}^h)\varepsilon_3^h\right], \quad \tau_{13}^h = G_{13}^h \gamma_{13}^h \tag{2.107}$$

$$\sigma_3^h = \frac{E_3^h}{1-\Delta}\left[(1-v_{12}^h v_{21}^h)\varepsilon_3^h + (v_{23}^h + v_{21}^h v_{13}^h)\varepsilon_2^h + (v_{13}^h + v_{12}^h v_{23}^h)\varepsilon_1^h\right], \quad \tau_{12}^h = G_{12}^h \gamma_{12}^h$$

$$\Delta = v_{12}^h v_{21}^h + v_{13}^h v_{31}^h + v_{23}^h v_{32}^h + v_{12}^h v_{23}^h v_{31}^h + v_{13}^h v_{21}^h v_{32}^h$$

2.3.4.2 Second Pass—2D Unit Cell Homogenization

After cores I and II are homogenized in x_2 as done in the first pass, a 3D local problem is thus simplified into a 2D problem, as shown in Figure 2.18.

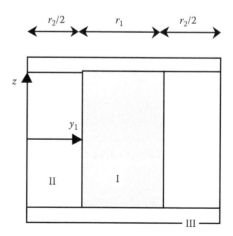

FIGURE 2.18
Second pass of MPH technique.

There are three regions, i.e., core I, core II, and two skin faces III, forming a three-phase homogenization problem with governing equations (2.98) through (2.103). The material properties of regions I and II are orthotropic and region III isotropic, which henceforth are expressed by engineering constants, Young's modulus E, shear modulus G, and Poisson's ratio v. Note for sandwich panels the following analyses in plane y_1-z are of plane strain deformations.

2.3.4.2.1 Homogenized Transverse Shear Stiffness

Denote the period of the honeycomb as $l_0 = a + b\cos\theta$ (Figure 2.12), and define the ratio $r_1 = b\cos\theta / l_0$, $r_2 = a / l_0$, $0 < r_1, r_2 < 1$, $r_1 + r_2 = 1$. Here h_1 for hexagons equals l_0 / δ. From (2.98) and (2.107) with $\varepsilon_2^h = 0$, the field equations for regions I and II may be given as

$$G_{13}^{hc\alpha}(\frac{\partial^2 \Pi_1^{[\alpha]}}{\partial z^2} + \frac{1}{h_1}\frac{\partial^2 \Pi_3^{[\alpha]}}{\partial y_1 \partial z}) +$$

$$\frac{E_1^{hc\alpha}}{1 - \Delta^{[\alpha]}}\left[(v_{31}^{hc\alpha} + v_{21}^{hc\alpha}v_{32}^{hc\alpha})\frac{1}{h_1}\frac{\partial^2 \Pi_3^{[\alpha]}}{\partial y_1 \partial z} + (1 - v_{32}^{hc\alpha}v_{23}^{hc\alpha})\frac{1}{h_1^2}\frac{\partial^2 \Pi_1^{[\alpha]}}{\partial y_1^2}\right] = 0$$

$$G_{13}^{hc\alpha}(\frac{1}{h_1}\frac{\partial^2 \Pi_1^{[\alpha]}}{\partial y_1 \partial z} + \frac{1}{h_1^2}\frac{\partial^2 \Pi_3^{[\alpha]}}{\partial y_1^2}) + \qquad\qquad (2.108)$$

$$\frac{E_3^{hc\alpha}}{1 - \Delta^{[\alpha]}}\left[(v_{13}^{hc\alpha} + v_{12}^{hc\alpha}v_{23}^{hc\alpha})\frac{1}{h_1}\frac{\partial^2 \Pi_1^{[\alpha]}}{\partial y_1 \partial z} + (1 - v_{12}^{hc\alpha}v_{21}^{hc\alpha})\frac{\partial^2 \Pi_3^{[\alpha]}}{\partial z^2}\right] = 0$$

$$\Delta^{[\alpha]} = v_{12}^{hc\alpha}v_{21}^{hc\alpha} + v_{13}^{hc\alpha}v_{31}^{hc\alpha} + v_{23}^{hc\alpha}v_{32}^{hc\alpha} + v_{12}^{hc\alpha}v_{23}^{hc\alpha}v_{31}^{hc\alpha} + v_{13}^{hc\alpha}v_{21}^{hc\alpha}v_{32}^{hc\alpha}$$

where α denotes region I or II.

The left and right boundary conditions for regions I and II are the continuity and the periodicity conditions for both stress and displacement fields. The top and bottom boundary conditions are the interactions between core and skin faces, which are involved with faces' internal fields. To simplify them, the approximation is made that faces are assumed infinite rigid, as those of Kelsey et al. (1958) and Penzien and Didriksson (1964), by which becomes zero at the top and bottom interfaces. Among all admissible displacement fields satisfying the above conditions, the following one in Fourier series is given based on the symmetric material and the antisymmetric loading about the panel middle plane:

$$\Pi_1^{[1]} = 0$$

$$\Pi_3^{[1]} = \sum_n a_n^{[1]} \sinh[\lambda_n^{[1]} h_1(y_1 - \frac{1}{2})]\cos(n\pi z), \quad n = 1,3,5 \cdots$$

$$\Pi_1^{[2]} = 0$$ (2.109)

$$\Pi_3^{[2]} = \sum_n a_n^{[2]} \sinh(\lambda_n^{[2]} h_1 y_1)\cos(n\pi z), \quad n = 1,3,5 \cdots$$

Substituting (2.109) into the second equation of (2.108) results in eigenvalues

$$\lambda_n^{[1]} = n\pi \sqrt{\frac{E_3^{hc1}(1 - v_{12}^{hc1} v_{21}^{hc1})}{G_{13}^{hc1}(1 - \Delta^{[1]})}}$$

(2.110)

$$\lambda_n^{[2]} = n\pi \sqrt{\frac{E_3^{hc2}(1 - v_{12}^{hc2} v_{21}^{hc2})}{G_{13}^{hc2}(1 - \Delta^{[2]})}}$$

And the first equation of (2.107) is satisfied by the variational partition method in a weak form since

$$\sum_n \int_{y_1}^{1/2} \int_{-1/2} \cosh[\lambda_n^{[1]} h_1(y_1 - \frac{l_0}{2})]\sin(n\pi z) = 0, \quad n = 1,3,5 \cdots$$

(2.111)

$$\sum_n \int_{y_1}^{1/2} \int_{-1/2} \cosh(\lambda_n^{[2]} h_1 y_1)\sin(n\pi z) = 0, \quad n = 1,3,5 \cdots$$

To find the coefficients a_n, we impose displacement continuity conditions between regions I and II:

$$\Pi_3^{[1]} = \Pi_3^{[2]} \quad \text{at } y_1 = r_2/2$$ (2.112)

and the equilibrium condition of shear stress

$$G_{13}^{hc1}(1 + \frac{1}{h_1}\frac{\partial \Pi_3^{[1]}}{\partial y_1}) = G_{13}^{hc2}(1 + \frac{1}{h_1}\frac{\partial \Pi_3^{[2]}}{\partial y_1}) \quad \text{at } y_1 = r_2/2$$ (2.113)

Note the continuity condition of normal stress is satisfied variationally by (2.111). With conditions (2.112) and (2.113), the coefficients are obtained:

$$a_n^{[1]} = \frac{4(G_{13}^{hc1} - G_{13}^{hc2})\sinh(\lambda_n^{[1]} h_1 \frac{r_1}{2})}{n\pi\left[-G_{13}^{hc2}\lambda_n^{[2]}\cosh(\lambda_n^{[2]} h_1 \frac{r_1}{2})\sinh(\lambda_n^{[1]} \frac{h_1 r_2}{2}) + G_{13}^{hc1}\lambda_n^{[1]}\cosh(\lambda_n^{[1]} \frac{h_1 r_2}{2})\sinh(\lambda_n^{[2]} h_1 \frac{r_1}{2})\right]}$$

$$a_n^{[2]} = \frac{4(G_{13}^{hc2} - G_{13}^{hc1})\sinh(\lambda_n^{[1]} \frac{h_1 r_2}{2})}{n\pi\left[-G_{13}^{hc2}\lambda_n^{[2]}\cosh(\lambda_n^{[2]} h_1 \frac{r_1}{2})\sinh(\lambda_n^{[1]} \frac{h_1 r_2}{2}) + G_{13}^{hc1}\lambda_n^{[1]}\cosh(\lambda_n^{[1]} \frac{h_1 r_2}{2})\sinh(\lambda_n^{[2]} h_1 \frac{r_1}{2})\right]}$$

(2.114)

From (2.93), (2.96), (2.97), and (2.107), the homogenized transverse shear stiffness is expressed:

$$H_{55}^H = \delta\langle C_{1313}\rangle = \delta G_{13}^H, \quad \text{where}$$

(2.115)

$$G_{13}^H = 2\int_0^{r_2/2}\int_{-1/2}^{1/2} G_{13}^{hc2}(1 + \frac{1}{h_1}\frac{\partial\Pi_3^{[2]}}{\partial y_1})\,dz\,dy_1 + 2\int_{r_2/2}^{1/2}\int_{-1/2}^{1/2} G_{13}^{hc1}(1 + \frac{1}{h_1}\frac{\partial\Pi_3^{[1]}}{\partial y_1})\,dz\,dy_1$$

And the integration of (2.115) results:

$$G_{13}^H = \sum_n \frac{4(-1)^{\frac{n-1}{2}}}{n^2\pi^2}\left[G_{13}^{hc1}r_1 + G_{13}^{hc2}r_2 + a_n^{[2]}G_{13}^{hc1}\frac{n\pi}{h_1}Sinh(\frac{r_1}{2}h_1\lambda_n^{[1]}) + a_n^{[1]}G_{13}^{hc2}\frac{n\pi}{h_1}Sinh(\frac{r_2}{2}h_1\lambda_n^{[2]})\right]$$ (2.116)

The formula of (2.116) can be easily calculated with a symbolic mathematical program such as Mathematica or Maple. Hereby the numerical results are given in Table 2.18 for two configurations of hexagons in the function of core thickness ratio h_1.

For regular hexagons with $t_2 = 2t_1$, $a = b$, $\theta = 60°$, the ratio of $r_2 = 2/3$, and the isotropic material with a Poisson's ratio of 0.3, the intermediate equivalent properties of regions I and II are obtained from (2.104) and (2.105):

$$E_3^{hc1} = 5.2\ G_{13}^{hc1}, \quad G_{13}^{hc1} = \frac{t_1}{\sqrt{3}b}G^c$$

$$E_3^{hc2} = 5.2\ G_{13}^{hc1}, \quad G_{13}^{hc2} = 2\ G_{13}^{hc1}$$

(2.117)

$$\frac{1 - v_{12}^{hc2}v_{21}^{hc2}}{1 - \Delta^{[2]}} = 1.0989$$

$$\frac{1 - v_{12}^{hc1}v_{21}^{hc1}}{1 - \Delta^{[1]}} = 1.0571 \text{ for } \frac{t_1}{b} \leq 0.1$$

TABLE 2.18

Homogenized Transverse Shear Stiffness G_{13}^H (Unit: G_{13}^{hc1})

h_1	LB	1/100	1/20	1/10	1/8	1/6	1/5	1/4	1/3	1/2	1	2	10	UB
$r_2 = 2/3$	1.500	1.500	1.504	1.508	1.510	1.514	1.517	1.521	1.529	1.543 (1.544)	1.580 (1.584)	1.618	1.656	1.666
$r_2 = 0.536$	1.366	1.366	1.370	1.374	1.377	1.381	1.384	1.388	1.395	1.410	1.448	1.486	1.525	1.536

Note: The numbers based on FEA are given in parentheses.

Further consider a group of irregular hexagons with $t_2 = 2t_1$, $a = b$, $\theta = 30°$, and the ratio of $r_2 = 0.536$. The properties of regions I and II are the same as in (2.117), wherein the value of

$$\frac{1 - v_{12}^{hc1} v_{21}^{hc1}}{1 - \Delta^{[1]}}$$

does not change. The results of two groups are shown in Table 2.18 and graphically in Figures 2.19 and 2.20.

2.3.4.2.1.1 Remarks

a. The variational principle leads to the lower bound (LB) and upper bound (UB) of shear stiffness as derived first by Kelsey et al. (1958), and later by Gibson and Ashby (1988) and Shi and Tong (1995a). With the simple application of the Reuss model and Voigt model for the 2D unit cell in Figure 2.18, respectively, the formulas for UB and LB can be written as

$$G_{13}^H \big|_{UB} = r_1 G_{13}^{hc1} + r_2 G_{13}^{hc2}$$

$$G_{13}^H \big|_{LB} = \frac{G_{13}^{hc1} G_{13}^{hc2}}{r_1 G_{13}^{hc2} + r_2 G_{13}^{hc1}}$$

(2.118)

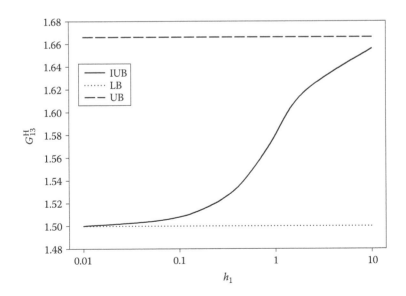

FIGURE 2.19
Transverse shear stiffness G_{13}^H with $r_2 = 2/3$.

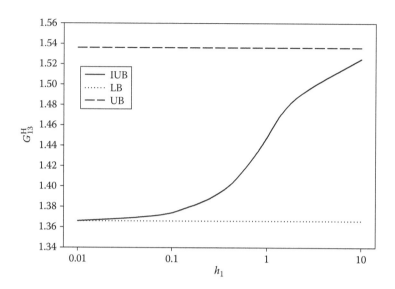

FIGURE 2.20
Transverse shear stiffness $G_{13}{}^H$ with $r_2 = 0.536$.

Note the two bounds are not functions of h_1. The substitution of (2.104), (2.105), into (2.118) leads to the formulas of LB and UB in functions of geometrical parameters.

$$G_{13}^H\Big|_{UB} = \frac{2t_1 b \cos^2\theta + t_2 a}{2b \sin\theta(a + b\cos\theta)} G_c$$

$$G_{13}^H\Big|_{LB} = \frac{b\cos\theta + a}{b\sin\theta(b/t_1 + 2a/t_2)} G_c$$

(2.119)

which are identical to those formulas in literature (Kelsey et al. 1958; Gibson and Ashby 1988).

b. Penzien and Didriksson (1964) gave a result similar to (2.116); however, its assumption of infinite Young's modulus in the y_1 axis is not right, which is corrected in this study by the GTM model and the weak form solution. Further, we should point out that the semianalytical approximation in Section 2.3.4.2.1 gives an improved upper bound (IUB) since skin faces are assumed infinite rigid. As shown in Figures 2.19 and 2.20, the IUB value converges to the two bounds when $h_1 \to 0$ or ∞, which confirms the solution.

The FEA results in Section 2.3.5 have two cases ($h_1 = 1$ and ½) illogically a little higher than IUB values, as modified in Table 2.18. The reason is attributed to the inconsistency between the IUB displacement field (2.109) and FEA modeling on the rigid assumption of skin-core joints. This influence, however, is quite small, and therefore IUB approximation of (2.116) is recommended as an effective formula that can be further combined with Equation (2.161) for design and optimization. For common hexagons that $h_1 < 1$, the LB value can be conservatively used in preliminary design. For refined sandwich analysis, such as FEA modeling, the more accurate shear stiffness is necessary; thus (2.116) and (2.161) or the unit cell FEA approach can be useful.

c. When skin effect is not taken into account (i.e., there is no normal constraint at the top and bottom boundaries), with continuity conditions (2.112) and (2.113) the exact solution of (2.108) may be simply known as

$$\Pi_1^{[1]} = 0$$

$$\Pi_3^{[1]} = \frac{h_1(G_{13}^{hc2} - G_{13}^{hc1})}{\frac{r_2}{r_1}G_{13}^{hc2} + G_{13}^{hc1}} y_1 + constant$$

$$\Pi_1^{[2]} = 0 \qquad\qquad (2.120)$$

$$\Pi_3^{[2]} = \frac{h_1(G_{13}^{hc1} - G_{13}^{hc2})}{G_{13}^{hc2} + \frac{r_1}{r_2}G_{13}^{hc1}} y_1$$

It can be easily checked that the substitution of (2.120) into (2.115) results in the identical expression of (2.119), which in turn validates the IUB approach.

d. From the GTM model in Section 2.3.4.1, the other homogenized transverse shear stiffness can be derived with the parallel model or Reuss model:

$$G_{23}^H = r_1 G_{23}^{hc1} + r_2 G_{23}^{hc2} \qquad\qquad (2.121)$$

and the substitution of (2.104) and (2.105) into (2.121) gives

$$G_{23}^H = \frac{b\sin\theta}{b\cos\theta + a}\frac{t_1}{b}G_c \qquad\qquad (2.122)$$

Clearly there is no skin effect for the case of G_{23}^H, which can also be intuitively seen from the GTM model in Figure 2.17.

e. For common honeycomb sandwiches, the thickness of skin faces is small relative to that of the core, while the shear modulus of the former is much higher. Owing to this fact, the equivalent shear stiffness of the sandwiches is overwhelmingly determined by core properties.

To further take the thickness of skin faces t_f into account, the unit width transverse shear stiffness of a sandwich panel can be written as per Allen (1969):

$$H_{55} = \frac{(t_f + \delta)^2}{\delta} G_{13}^H$$

$$H_{44} = \frac{(t_f + \delta)^2}{\delta} G_{23}^H$$

(2.123)

2.3.4.2.2 Homogenized In-Plane Stretch Stiffness

The governing field equations and the boundary conditions are the same as (2.108) and those of Section 2.3.4.2.1:

$$G_{13}^{hc\alpha} \left(\frac{\partial^2 U_1^{[\alpha]}}{\partial z^2} + \frac{1}{h_1} \frac{\partial^2 U_3^{[\alpha]}}{\partial y_1 \partial z} \right) +$$

$$\frac{E_1^{hc\alpha}}{1 - \Delta^{[\alpha]}} \left[(v_{31}^{hc\alpha} + v_{21}^{hc\alpha} v_{32}^{hc\alpha}) \frac{1}{h_1} \frac{\partial^2 U_3^{[\alpha]}}{\partial y_1 \partial z} + (1 - v_{32}^{hc\alpha} v_{23}^{hc\alpha}) \frac{1}{h_1^2} \frac{\partial^2 U_1^{[\alpha]}}{\partial y_1^2} \right] = 0$$

$$G_{13}^{hc\alpha} \left(\frac{1}{h_1} \frac{\partial^2 U_1^{[\alpha]}}{\partial y_1 \partial z} + \frac{1}{h_1^2} \frac{\partial^2 U_3^{[\alpha]}}{\partial y_1^2} \right) +$$

(2.124)

$$\frac{E_3^{hc\alpha}}{1 - \Delta^{[\alpha]}} \left[(v_{13}^{hc\alpha} + v_{12}^{hc\alpha} v_{23}^{hc\alpha}) \frac{1}{h_1} \frac{\partial^2 U_1^{[\alpha]}}{\partial y_1 \partial z} + (1 - v_{12}^{hc\alpha} v_{21}^{hc\alpha}) \frac{\partial^2 U_3^{[\alpha]}}{\partial z^2} \right] = 0$$

$$\Delta^{[\alpha]} = v_{12}^{hc\alpha} v_{21}^{hc\alpha} + v_{13}^{hc\alpha} v_{31}^{hc\alpha} + v_{23}^{hc\alpha} v_{32}^{hc\alpha} + v_{12}^{hc\alpha} v_{23}^{hc\alpha} v_{31}^{hc\alpha} + v_{13}^{hc\alpha} v_{21}^{hc\alpha} v_{32}^{hc\alpha}$$

Due to the loading condition, the solution approximation is made contrary to the case in Section 2.3.4.2.1; i.e., in this case the displacement is ensured to satisfy governing equations strictly in y_1 and weakly in z. By the double symmetry of material and loading, the displacement field is constructed as

$$U_1^{[1]} = \sum_n b_n^{[1]} \sinh[\lambda_n^{[1]} h_1 (y_1 - \frac{1}{2})] \cos(n\pi z), \quad n = 1,3,5 \cdots$$

$$U_3^{[1]} = \sum_n c_n \sin(n\pi z), \quad n = 1,3,5 \cdots$$

(2.125)

$$U_1^{[2]} = \sum_n b_n^{[2]} \sinh(\lambda_n^{[2]} h_1 y_1) \cos(n\pi z), \quad n = 1,3,5 \cdots$$

$$U_3^{[2]} = \sum_n c_n \sin(n\pi z), \quad n = 1,3,5 \cdots$$

The substitution of (2.125) into the first equation of (2.124) results in eigenvalues

$$\lambda_n^{[1]} = n\pi \sqrt{\frac{G_{13}^{hc1}(1-\Delta^{[1]})}{E_1^{hc1}(1-v_{23}^{hc1}v_{32}^{hc1})}}$$

$$\lambda_n^{[2]} = n\pi \sqrt{\frac{G_{13}^{hc2}(1-\Delta^{[2]})}{E_1^{hc2}(1-v_{23}^{hc2}v_{32}^{hc2})}}$$

(2.126)

Note in (2.125) there are three unknown coefficients, and the second equation of (2.124) cannot be simply satisfied by the partition method in weak form as (2.111). An effective approach is to follow the Rayleigh–Ritz method that is convenient to treat orthonormal series. The quadratic energy functional of (2.124) is given by

$$I(U) = Q_s + Q_y + Q_z$$

$$Q_s = 2\int_{-\frac{1}{2}}^{\frac{1}{2}}\int_0^{\frac{r_2}{2}} \frac{1}{2} G_{13}^{hc2} (\frac{\partial U_1^{[2]}}{\partial z} + \frac{1}{h_1}\frac{\partial U_3^{[2]}}{\partial y_1})^2 dy_1 dz$$

$$+2\int_{-\frac{1}{2}}^{\frac{1}{2}}\int_{\frac{r_2}{2}}^{\frac{1}{2}} \frac{1}{2} G_{13}^{hc1} (\frac{\partial U_1^{[1]}}{\partial z} + \frac{1}{h_1}\frac{\partial U_3^{[1]}}{\partial y_1})^2 dy_1 dz$$

$$Q_y = 2\int_{-\frac{1}{2}}^{\frac{1}{2}}\int_0^{\frac{r_2}{2}} \frac{1}{2}\frac{E_1^{hc2}}{1-\Delta^{[2]}}\left[(v_{31}^{hc2}+v_{21}^{hc2}v_{32}^{hc2})\frac{\partial U_1^{[2]}}{\partial z}+(1-v_{23}^{hc2}v_{32}^{hc2})(\frac{1}{h_1}\frac{\partial U_1^{[2]}}{\partial y_1}+1)\right]\frac{1}{h_1}\frac{\partial U_1^{[2]}}{\partial y_1}+1)dy_1 dz$$

$$+2\int_{-\frac{1}{2}}^{\frac{1}{2}}\int_{\frac{r_2}{2}}^{\frac{1}{2}} \frac{1}{2}\frac{E_1^{hc1}}{1-\Delta^{[1]}}\left[(v_{31}^{hc1}+v_{21}^{hc1}v_{32}^{hc1})\frac{\partial U_3^{[1]}}{\partial z}+(1-v_{23}^{hc1}v_{32}^{hc1})\frac{1}{h_1}\frac{\partial U_1^{[1]}}{\partial y_1}+1\right](\frac{1}{h_1}\frac{\partial U_1^{[1]}}{\partial y_1}+1)dy_1 dz$$

(2.127)

$$Q_z = 2\int_{-\frac{1}{2}}^{\frac{1}{2}}\int_0^{\frac{r_2}{2}} \frac{1}{2}\frac{E_3^{hc2}}{1-\Delta^{[2]}}\left[(v_{13}^{hc2}+v_{12}^{hc2}v_{23}^{hc2})(\frac{1}{h_1}\frac{\partial U_1^{[2]}}{\partial y_1}+1)+(1-v_{21}^{hc2}v_{12}^{hc2})\frac{\partial U_3^{[2]}}{\partial z}\right]\frac{\partial U_3^{[2]}}{\partial z}dy_1 dz$$

$$+2\int_{-\frac{1}{2}}^{\frac{1}{2}}\int_{\frac{r_2}{2}}^{\frac{1}{2}} \frac{1}{2}\frac{E_3^{hc1}}{1-\Delta^{[1]}}\left[(v_{13}^{hc1}+v_{12}^{hc1}v_{23}^{hc1})(\frac{1}{h_1}\frac{\partial U_1^{[1]}}{\partial y_1}+1)+(1-v_{21}^{hc1}v_{12}^{hc1})\frac{\partial U_3^{[1]}}{\partial z}\right]\frac{\partial U_3^{[1]}}{\partial z}dy_1 dz$$

where Q_s, Q_y, and Q_z denote strain energy in shear, y_1-normal, and z-normal, respectively.

The three equations for three unknown coefficients are the displacement continuity condition between region I and II:

$$U_1^{[1]} = U_1^{[2]} \quad \text{at } y_1 = r_2/2$$

(2.128)

and the differentiation of functional $I(U)$ about any two unknowns, say two b_n:

$$\frac{\partial I(U)}{\partial b_n^{[1]}} = 0$$

$$\frac{\partial I(U)}{\partial b_n^{[2]}} = 0 \tag{2.129}$$

The substitution of (2.125) and (2.127) into (2.128) and (2.129) solves the three coefficients, where the symbolic calculation can be conveniently carried out by Mathematica or Maple. The results of the coefficients b_n and c_n are omitted here due to their lengthy symbolic expressions.

From (2.85), (2.96), (2.97), and (2.107), the homogenized in-plane stretch stiffness is given:

$$A_{11}^H = \delta \left\langle C_{1111} \right\rangle = \delta E_{11}^H$$

where

$$E_{11}^H = 2 \int_0^{r_2/2} \int_{-1/2}^{1/2} \frac{E_1^{hc2}}{1-\Delta^{[2]}} \left[(1-v_{23}^{hc2}v_{32}^{hc2})(1+\frac{1}{h_1}\frac{\partial U_1^{[2]}}{\partial y_1}) + (v_{31}^{hc2}+v_{21}^{hc2}v_{32}^{hc2})\frac{\partial U_3^{[2]}}{\partial z} \right] dz\, dy_1 \tag{2.130}$$

$$+2 \int_{r_2/2}^{1/2} \int_{-1/2}^{1/2} \frac{E_1^{hc1}}{1-\Delta^{[1]}} \left[(1-v_{23}^{hc1}v_{32}^{hc1})(1+\frac{1}{h_1}\frac{\partial U_1^{[1]}}{\partial y_1}) + (v_{31}^{hc1}+v_{21}^{hc1}v_{32}^{hc1})\frac{\partial U_3^{[1]}}{\partial z} \right] dz\, dy_1$$

And the integration of (2.130) results in

$$E_{11}^H = \sum_n 2(-1)^{\frac{n+3}{2}} \left\{ \frac{c_n(v_{31}^{hc1}+v_{21}^{hc1}v_{32}^{hc1}) + [\frac{2b_n^{[1]}}{n\pi}\sinh(\lambda_n^{[1]}\frac{h_1 r_1}{2}) - \frac{4}{n^2\pi^2}](1-v_{23}^{hc1}v_{32}^{hc1})}{1-\Delta^{[1]}} r_1 E_1^{hc1} + \right.$$

$$\left. \frac{c_n(v_{31}^{hc2}+v_{21}^{hc2}v_{32}^{hc2}) + [\frac{2b_n^{[2]}}{n\pi}\sinh(\lambda_n^{[2]}\frac{h_1 r_2}{2}) - \frac{4}{n^2\pi^2}](1-v_{23}^{hc2}v_{32}^{hc2})}{1-\Delta^{[2]}} r_2 E_1^{hc2} \right\} \quad n=1,3,5\cdots \tag{2.131}$$

Let all the geometrical properties be the same as those of Section 2.3.4.2.1, with $r_2 = 2/3$, then the intermediate equivalent properties expressed by E_1^{hc2} are

$$E_1^{hc2} = \frac{2t_1}{\sqrt{3b}} E_c$$

$$G_{13}^{hc1} = \frac{1}{5.2} E_1^{hc2} \qquad G_{13}^{hc2} = \frac{1}{2.6} E_1^{hc2} \tag{2.132}$$

$$1-\Delta^{[2]} = 0.91$$

$\frac{t_1}{b}$	0.1	0.05	0.025	0.01	(unit: E_1^{hc2})
$\dfrac{(1-v_{32}^{hc1}v_{23}^{hc1})E_1^{hc1}}{1-\Delta^{[1]}}$	0.135833	0.133059	0.132365	0.132171	
$\dfrac{(v_{31}^{hc1}+v_{21}^{hc1}v_{32}^{hc1})E_1^{hc1}}{1-\Delta^{[1]}}$	0.122376	0.122700	0.122781	0.122804	(2.133)
$\dfrac{(1-v_{32}^{hc1}v_{23}^{hc1})E_3^{hc1}}{1-\Delta^{[1]}}$	2.114191	2.114153	2.114144	2.114144	
$\dfrac{(v_{31}^{hc1}+v_{21}^{hc1}v_{32}^{hc1})E_3^{hc1}}{1-\Delta^{[1]}}$	0.122376	0.122700	0.122781	0.122804	

The substitution of (2.132) and (2.133) into (2.131) results in the homogenized stretch stiffness in terms of thickness ratio h_1, and the numerical data are given in Table 2.19 and shown in Figure 2.21.

2.3.4.2.3 Remarks

a. With Becker's (1998) investigation of thickness effect on honeycomb in-plane stiffness, there seems to be no further attentions on its influence on honeycomb sandwich computations. Actually, almost

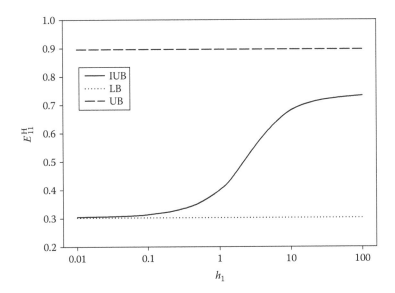

FIGURE 2.21
Stretch stiffness E_{11}^H with $t_1/b = 0.05$.

TABLE 2.19

Homogenized In-Plane Stretch Stiffness E_{11}^H with $r_2 = 2/3$ (Unit: E_1^{hc2})

h_1	LB	1/100	1/20	1/10	1/8	1/6	1/5	1/4	1/3	1/2	1	2	10	100	10,000	UB
$t_1/b = 0.1$		0.310	0.313	0.318	0.320	0.324	0.327	0.331	0.340	0.355	0.400	0.484	0.674	0.725	0.732	0.898
$t_1/b = 0.05$		0.304	0.308	0.313	0.315	0.319	0.322	0.324	0.334	0.349	0.496	0.481	0.672	0.725		
$t_1/b = 0.025$	0.302	0.302	0.307	0.311	0.314	0.317	0.321	0.325	0.333	0.348	0.395	0.481	0.672	0.724	0.730	0.896
$t_1/b = 0.01$			0.306	0.311	0.313		0.320				0.394	0.480				

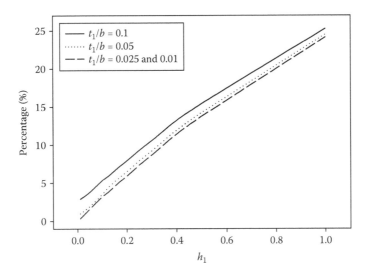

FIGURE 2.22
Underestimate percentage of $E_{11}{}^H$.

all of today's sandwich computations follow the three-layer theories where equivalent core properties are obtained without skin effect considered. The conventional approach just has its value corresponding to the LB, either implicated in FEA modeling or explicated in the formulas. As shown in Figure 2.22, this may result in stretch stiffness underestimated by about 3 to 25% when h_1 is increased from 0.1 to 1.0, the range of common honeycomb applications. It is also found that the cell wall thickness ratio t_1/b has little influence on skin effect, particularly when its value is less than 0.05.

b. By applying the Reuss model and Voigt model in this case, the formulas of LB and UB are

$$E_{11}^H\Big|_{UB} = r_1 \frac{(1-v_{23}^{hc1}v_{32}^{hc1})E_1^{hc1}}{1-\Delta^{[1]}} + r_2 \frac{(1-v_{23}^{hc2}v_{32}^{hc2})E_1^{hc2}}{1-\Delta^{[2]}}$$

$$E_{11}^H\Big|_{LB} = \frac{\sigma_1^h}{r_1(\frac{\sigma_1^h}{E_1^{hc1}} - v_{21}^{hc1}\frac{\sigma_2^{hc1}}{E_2^{hc1}}) + r_2(\frac{\sigma_1^h}{E_1^{hc2}} - v_{21}^{hc2}\frac{\sigma_2^{hc2}}{E_2^{hc2}})}$$

(2.134)

Note that to be consistent with the stationary variational principle, in (2.134) Poisson's effect along the z axis is considered for UB but not for LB (i.e., $\varepsilon_2^h = \varepsilon_3^h = 0$ and $\varepsilon_2^h = \sigma_3^h = 0$ for UB and LB, respectively). The substitution of (2.104) and (2.105) into (2.134) finally results in

$$E_{11}^H\big|_{UB} = \frac{\dfrac{1}{1-v_c^2} + \dfrac{2bt_1}{at_2}\cos^4\theta\,\Phi}{(\cos\theta\, b/a+1)\sin\theta} - (t_2/2b)E_c$$

$$E_{11}^H\big|_{LB} = \frac{(\cos\theta\, b/a+1)}{\dfrac{t_2\, b\sin\theta}{2t_1a\cos^2\theta\,[1+tg^2\theta\,(t_1/b)^2]} + \sin\theta} - (t_2/2b)E_c \qquad (2.135)$$

$$\Phi_1 = \frac{1+tg^2\theta\,(t_1/b)^2\,[1-\sin^2\theta v_c^2]}{1-\dfrac{3}{4}v_c^2 - \dfrac{1}{4}v_c^2\cos(4\theta)}$$

where Poisson's effect is represented by the existence of v_c.

When the regular hexagon is considered with parameters $t_2 = 2t_1$, $a = b$, and $\theta = 60°$, then for small t_1/b, (2.135) is approximated as

$$E_{11}^H\big|_{LB} \approx \frac{\sqrt{3}}{5}(t_1/b)E_c$$

$$\qquad (2.136)$$

$$E_{11}^H\big|_{UB} \approx \frac{\sqrt{3}}{2(1-v_c^2)}(t_1/b)E_c$$

which agree with the results of

$$\frac{\sqrt{3}}{5}(t_1/b)E_c$$

given by Shi and Tong (1995b) and Becker (1998), whereas the formula of Parton and Kudryavtsev (1993) seems to overestimate stretch stiffness.

The formula (2.131) is an improved upper bound (IUB) because the skin faces are assumed infinite rigid. Note the strain is allowed in three directions so that the derivation is consistent with real situations such as in compression buckling. It can be seen that when $h_1 \to 0$, the IUB of this study converges to the LB in (2.135) when the cell wall thickness ratio is not higher than 0.025. The IUB does not converge to the UB of (2.135) when $h_1 \to \infty$ because the UB formula in (2.135) assumes zero strain in the three directions.

c. Another homogenized in-plane stiffness E_{21}^H is readily obtainable with the same displacement field (2.124)

$A_{21}^H = \delta \langle C_{2211} \rangle = \delta E_{21}^H,$ where

$$E_{21}^H = 2 \int_0^{r_2/2} \int_{-1/2}^{1/2} \frac{E_2^{hc2}}{1-\Delta^{[2]}} \left[(v_{12}^{hc2} + v_{13}^{hc2} v_{32}^{hc2})(1 + \frac{1}{h_1} \frac{\partial U_1^{[2]}}{\partial y_1}) + (v_{32}^{hc2} + v_{12}^{hc2} v_{31}^{hc2}) \frac{\partial U_3^{[2]}}{\partial z} \right] dx dy_1 \qquad (2.137)$$

$$+ 2 \int_{r_2/2}^{1/2} \int_{-1/2}^{1/2} \frac{E_2^{hc1}}{1-\Delta^{[1]}} \left[(v_{12}^{hc1} + v_{13}^{hc1} v_{32}^{hc1})(1 + \frac{1}{h_1} \frac{\partial U_1^{[1]}}{\partial y_1}) + (v_{32}^{hc1} + v_{12}^{hc1} v_{31}^{hc1}) \frac{\partial U_3^{[1]}}{\partial z} \right] dz dy_1$$

By considering $E_2^{hc2} = 0$ in (2.104), the integration of (2.137) results:

$$E_{21}^H = \sum_n 2(-1)^{\frac{n+3}{2}} \frac{c_n (v_{32}^{hc1} + v_{12}^{hc1} v_{31}^{hc1}) + [\frac{2b_n^{[1]}}{n\pi} \sinh(\lambda_n^{[1]} \frac{h_1 r_1}{2}) - \frac{4}{n^2 \pi^2}](v_{12}^{hc1} + v_{13}^{hc1} v_{32}^{hc1})}{1-\Delta^{[1]}} r_1 E_2^{hc1} \qquad (2.138)$$

$n = 1,3,5\cdots$

The eigenvalues and coefficients in (2.138) are the same as those of (2.131).

In converse to the case of E_{11}^H, the formula (2.138) is an improved lower bound (ILB) with the assumption of infinite rigid skin faces, which is also noted by Hohe and Becker (2001). The numerical results are omitted here.

d. Similar to the case of G_{23}^H, from the GTM model in Section 2.3.4.1, the third homogenized in-plane stiffness E_{22}^H can be easily derived with the parallel model or Reuss model considering $\varepsilon_3^h = 0$:

$$E_{22}^H = r_1 (1 - v_{13}^{hc1} v_{31}^{hc1}) \frac{E_2^{hc1}}{1-\Delta^{[1]}} + r_2 (1 - v_{13}^{hc2} v_{31}^{hc2}) \frac{E_2^{hc2}}{1-\Delta^{[2]}} \qquad (2.139)$$

And the substitution of (2.104) and (2.105) into (2.139) finally gives

$$E_{22}^H = \frac{\sin^3 \theta \; \Phi_2}{\cos \theta + a/b} \left(\frac{t_1}{b} \right) E_c$$

$$\Phi_2 = \frac{1 + (t_1/b)^2 [\csc \theta \; ctg^2\theta - v_c^2 \cos^2 \theta \csc^2 \theta]}{1 - \frac{3}{4} v_c^2 - \frac{1}{4} v_c^2 \cos(4\theta)} \qquad (2.140)$$

For regular hexagons with $t_2 = 2t_1$, $a = b$, $\theta = 60°$, and small t_1/b, (2.140) is approximated as

$$E_{22}^H \approx \frac{\sqrt{3}}{4(1 - \frac{5}{8} v_c^2)} (t_1/b) E_c \qquad (2.141)$$

The real value of E_{22}^H should be between zero and the full Poisson's effect, i.e., between

$$\frac{\sqrt{3}}{4}(t_1/b)E_c \text{ and } \frac{\sqrt{3}}{4(1-\frac{5}{8}v_c^2)}(t_1/b)E_c.$$

Compared with (2.136), it can be found that the stretch stiffness becomes little anisotropic due to skin effect.

e. The unit width stretch stiffness of a honeycomb sandwich thus can be calculated with

$$A_{11} = \delta E_{11}^H + 2t_f \frac{E_f}{1-v_f^2}$$

$$A_{22} = \delta E_{22}^H + 2t_f \frac{E_f}{1-v_f^2} \tag{2.142}$$

$$A_{21} = \delta E_{21}^H + 2t_f \frac{v_f E_f}{1-v_f^2}$$

2.3.4.2.4 Homogenized Flexural Stiffness

Similar to Sections 2.3.4.2.1 and 2.3.4.2.2, from (2.102) the governing equations of cylindrical bending are given below:

$$G_{13}^{hc\alpha}(\frac{\partial^2 V_1^{[\alpha]}}{\partial z^2} + \frac{1}{h_1}\frac{\partial^2 V_3^{[\alpha]}}{\partial y_1 \partial z})$$

$$+ \frac{E_1^{hc\alpha}}{1-\Delta^{[\alpha]}}\left[(v_{31}^{hc\alpha} + v_{21}^{hc\alpha}v_{32}^{hc\alpha})\frac{1}{h_1}\frac{\partial^2 V_3^{[\alpha]}}{\partial y_1 \partial z} + (1-v_{32}^{hc\alpha}v_{23}^{hc\alpha})\frac{1}{h_1^2}\frac{\partial^2 V_1^{[\alpha]}}{\partial y_1^2}\right] = 0$$

$$G_{13}^{hc\alpha}(\frac{1}{h_1}\frac{\partial^2 V_1^{[\alpha]}}{\partial y_1 \partial z} + \frac{1}{h_1^2}\frac{\partial^2 V_3^{[\alpha]}}{\partial y_1^2}) \tag{2.143}$$

$$+ \frac{E_3^{hc\alpha}}{1-\Delta^{[\alpha]}}\left[(v_{13}^{hc\alpha} + v_{12}^{hc\alpha}v_{23}^{hc\alpha})\frac{1}{h_1}\frac{\partial^2 V_1^{[\alpha]}}{\partial y_1 \partial z} + (1-v_{12}^{hc\alpha}v_{21}^{hc\alpha})\frac{\partial^2 V_3^{[\alpha]}}{\partial z^2}\right] = 0$$

$$\Delta^{[\alpha]} = v_{12}^{hc\alpha}v_{21}^{hc\alpha} + v_{13}^{hc\alpha}v_{31}^{hc\alpha} + v_{23}^{hc\alpha}v_{32}^{hc\alpha} + v_{12}^{hc\alpha}v_{23}^{hc\alpha}v_{31}^{hc\alpha} + v_{13}^{hc\alpha}v_{21}^{hc\alpha}v_{32}^{hc\alpha}$$

Based on the symmetry of material and loading about the z axis, and the antisymmetry of loading about the y_1 axis, the displacement field is constructed as

$$V_1^{[1]} = \sum_n d_n^{[1]} \sinh[\lambda_n^{[1]} h_1(y_1 - \frac{1}{2})]\sin(n\pi z), \quad n = 2,4,6\cdots$$

$$V_3^{[1]} = \frac{1}{2} v_{13}^{hc1} z^2 + \sum_n f_n \cos(n\pi z), \quad n = 2,4,6\cdots$$

$$V_1^{[2]} = \sum_n d_n^{[2]} \sinh(\lambda_n^{[2]} h_1 y_1)\sin(n\pi z), \quad n = 2,4,6\cdots$$

(2.144)

$$V_3^{[2]} = \frac{1}{2} v_{13}^{hc2} z^2 + \sum_n f_n \cos(n\pi z), \quad n = 2,4,6\cdots$$

Note in (2.144) the first term of V_3 expression is added to ensure displacement continuity. As the equilibrium condition along the y_1 axis is more important than that along the z axis, (2.144) is substituted into the first equation of (2.143) to obtain eigenvalues

$$\lambda_n^{[1]} = n\pi \sqrt{\frac{G_{13}^{hc1}(1-\Delta^{[1]})}{E_1^{hc1}(1-v_{23}^{hc1}v_{32}^{hc1})}}$$

(2.145)

$$\lambda_n^{[2]} = n\pi \sqrt{\frac{G_{13}^{hc2}(1-\Delta^{[2]})}{E_1^{hc2}(1-v_{23}^{hc2}v_{32}^{hc2})}}$$

The solving of three unknown coefficients in (2.144) just repeats those done in Section 2.3.4.2.2 by the Rayleigh–Ritz method. The expression of strain energy is given below, while the symbolic calculation and expressions of d_n and f_n are omitted here.

$$I(V) = Q_s + Q_y + Q_z$$

$$Q_s = 2\int_{-\frac{1}{2}}^{\frac{1}{2}}\int_0^{\frac{r_2}{2}} \frac{1}{2} G_{13}^{hc2} (\frac{\partial V_1^{[2]}}{\partial z} + \frac{1}{h_1}\frac{\partial V_3^{[2]}}{\partial y_1})^2 dy_1 dz$$

$$+ 2\int_{-\frac{1}{2}}^{\frac{1}{2}}\int_{\frac{r_2}{2}}^{\frac{1}{2}} \frac{1}{2} G_{13}^{hc1} (\frac{\partial V_1^{[1]}}{\partial z} + \frac{1}{h_1}\frac{\partial V_3^{[1]}}{\partial y_1})^2 dy_1 dz$$

$$Q_y = 2\int_{-\frac{1}{2}}^{\frac{1}{2}}\int_0^{\frac{r_2}{2}} \frac{1}{2}\frac{E_1^{hc2}}{1-\Delta^{[2]}}\left[(v_{31}^{hc2}+v_{21}^{hc2}v_{32}^{hc2})\frac{\partial V_3^{[2]}}{\partial z} + (1-v_{32}^{hc2}v_{23}^{hc2})(\frac{1}{h_1}\frac{\partial V_1^{[2]}}{\partial y_1}+z)\right](\frac{1}{h_1}\frac{\partial V_1^{[2]}}{\partial y_1}+z)dy_1 dz$$

$$+ 2\int_{-\frac{1}{2}}^{\frac{1}{2}}\int_{\frac{r_2}{2}}^{\frac{1}{2}} \frac{1}{2}\frac{E_1^{hc1}}{1-\Delta^{[1]}}\left[(v_{31}^{hc1}+v_{21}^{hc1}v_{32}^{hc1})\frac{\partial V_3^{[1]}}{\partial z} + (1-v_{32}^{hc1}v_{23}^{hc1})(\frac{1}{h_1}\frac{\partial V_1^{[1]}}{\partial y_1}+z)\right](\frac{1}{h_1}\frac{\partial V_1^{[1]}}{\partial y_1}+z)dy_1 dz \qquad (2.146)$$

$$Q_z = 2\int_{-\frac{1}{2}}^{\frac{1}{2}}\int_0^{\frac{r_2}{2}} \frac{1}{2}\frac{E_3^{hc2}}{1-\Delta^{[2]}}\left\{(v_{13}^{hc2}+v_{13}^{hc2}v_{23}^{hc2})(\frac{1}{h_1}\frac{\partial V_1^{[2]}}{\partial y_1}+z) + [\frac{\partial V_3^{[2]}}{\partial z} - (v_{13}^{hc2}+v_{12}^{hc2}v_{23}^{hc2})z](1-v_{12}^{hc2}v_{21}^{hc2})\right\}\times$$

$$[\frac{\partial V_3^{[2]}}{\partial z} - (v_{13}^{hc2}+v_{12}^{hc2}v_{23}^{hc2})z]dy_1 dz +$$

$$2\int_{-\frac{1}{2}}^{\frac{1}{2}}\int_{\frac{r_2}{2}}^{\frac{1}{2}} \frac{1}{2}\frac{E_3^{hc1}}{1-\Delta^{[1]}}\left\{(v_{13}^{hc1}+v_{13}^{hc1}v_{23}^{hc1})(\frac{1}{h_1}\frac{\partial V_1^{[1]}}{\partial y_1}+z) + [\frac{\partial V_3^{[1]}}{\partial z} - (v_{13}^{hc1}+v_{12}^{hc1}v_{23}^{hc1})z](1-v_{12}^{hc1}v_{21}^{hc1})\right\}\times$$

$$[\frac{\partial V_3^{[1]}}{\partial z} - (v_{13}^{hc1}+v_{12}^{hc1}v_{23}^{hc1})z]dy_1 dz$$

From (2.85), (2.96), (2.97), and (2.107), we can have homogenized flexural stiffness as

$$D_{11}^H = \delta^3 \langle C_{1111}^* \rangle = \frac{\delta^3}{12} E_{11}^{*H}, \quad \text{where}$$

$$E_{11}^{*H} = 12\times\left\{ 2\int_0^{r_2/2}\int_{-1/2}^{1/2} \frac{E_1^{hc2}}{1-\Delta^{[2]}}\left[(1-v_{23}^{hc2}v_{32}^{hc2})(z^2+z\frac{1}{h_1}\frac{\partial V_1^{[2]}}{\partial y_1}) + z(v_{31}^{hc2}+v_{21}^{hc2}v_{32}^{hc2})\frac{\partial V_3^{[2]}}{\partial z}\right]dz\,dy_1 \right. \qquad (2.147)$$

$$\left. + 2\int_{r_2/2}^{1/2}\int_{-1/2}^{1/2} \frac{E_1^{hc1}}{1-\Delta^{[1]}}\left[(1-v_{23}^{hc1}v_{32}^{hc1})(z^2+z\frac{1}{h_1}\frac{\partial V_1^{[1]}}{\partial y_1}) + z(v_{31}^{hc1}+v_{21}^{hc1}v_{32}^{hc1})\frac{\partial V_3^{[1]}}{\partial z}\right]dz\,dy_1 \right\}$$

The integration of (2.147) results in

$$E_{11}^{*H} = 12\times\sum_n (-1)^{\frac{n}{2}}\left\{ \frac{r_1 f_n(v_{31}^{hc1}+v_{21}^{hc1}v_{32}^{hc1}) - [\frac{2d_n^{[1]}}{h_1 n\pi}\sinh(\lambda_n^{[1]}\frac{h_1 r_1}{2}) + \frac{2(-1)^{\frac{n}{2}}r_1}{n^2\pi^2}](1-v_{23}^{hc1}v_{32}^{hc1})}{1-\Delta^{[1]}} E_1^{hc1} \right.$$

$$(2.148)$$

$$\left. + \frac{r_2 f_n(v_{31}^{hc2}+v_{21}^{hc2}v_{32}^{hc2}) - [\frac{2d_n^{[2]}}{h_1 n\pi}\sinh(\lambda_n^{[2]}\frac{h_1 r_2}{2}) + \frac{2(-1)^{\frac{n}{2}}r_2}{n^2\pi^2}](1-v_{23}^{hc1}v_{32}^{hc1})}{1-\Delta^{[2]}} E_1^{hc2} \right\} \quad n = 2,4,6\cdots$$

The alternative way is to use the form of energy equivalence $\left\langle E_{11}^{*H} \right\rangle = 12 \times 2 \times I(V)$.

With the hexagon configuration data as before, the numerical results for the flexural stiffness are given in Table 2.20 and shown in Figure 2.23.

2.3.4.2.4.1 Remarks

a. The flexural stiffness is distinguished from the stretch stiffness, i.e., $E_{11}^{*H} \neq E_{11}^{H}$. The difference can be as high as 25% for the case $h_1 = 1$, as shown in Figure 2.24. It is interesting to note that when $h_1 \to 0$ or ∞, the two stiffnesses converge to each other.

Most current sandwich computational approaches have the errors in the continuum modeling of the honeycomb core because skin effect has never been taken into account. For regular hexagons, as shown in Figure 2.25, the flexural stiffness is 5 to 40% underestimated when h_1 increases from 0.05 to 1.0. The consequence of the errors has to be evaluated in particular cases; however, it is suggested that all computational modeling use the corrected homogenized properties since the required efforts are minimum.

b. The formula of the panel unit width cylindrical flexural rigidity is

$$D_{11} = \frac{\delta^3}{12} E_{11}^{*H} + \frac{(t_f + \delta)^2 t_f}{2} \frac{E_f}{1 - v_f^2}$$

$$D_{22} = \frac{\delta^3}{12} E_{22}^{*H} + \frac{(t_f + \delta)^2 t_f}{2} \frac{E_f}{1 - v_f^2} \qquad (2.149)$$

$$D_{21} = \frac{\delta^3}{12} E_{21}^{*H} + \frac{(t_f + \delta)^2 t_f}{2} \frac{v_f E_f}{1 - v_f^2}$$

When t_1/b increases from 0.01 to 0.1, the hexagon core can have more than 20% contribution in total flexural rigidity if $\delta/t_f = 50$. The antiplane assumption should thus be carefully used in computations of honeycomb sandwiches. By using an equivalent antiplane assumption, the core shear stiffness was defined by Allen (1969) as

$$G_{13}^{*H} = \frac{G_{13}^{H}}{1 + \dfrac{E_{11}^{*H}(1 - v_f^2)}{6E_f} \dfrac{\delta^2}{t_f(\delta + t_f)}} \qquad (2.150)$$

It is found that the transverse shear stiffness can be much overestimated by ignoring the core's flexural stiffness. This may partially explain the long existing contradiction that shear stiffness obtained from testing is always larger than the theoretical upper bound,

TABLE 2.20

Homogenized Flexural Stiffness E_{11}^{*H} with $r_2 = 2/3$ (Unit: E_1^{hc2})

h_1	LB	1/10,000	1/100	1/20	1/10	1/8	1/6	1/5	1/4	1/3	1/2	1	2	10	100	10,000	UB
$t_1/b = 0.1$	0.304	0.308	0.310	0.320	0.334	0.340	0.351	0.359	0.371	0.391	0.427	0.514	0.608	0.705	0.727	0.730	0.898
$t_1/b = 0.05$	0.302	0.303	0.305	0.315	0.329	0.335	0.346	0.354	0.367	0.386	0.423	0.511	0.606	0.704	0.726	0.729	
$t_1/b = 0.025$	0.300	0.301	0.303	0.314	0.327	0.334	0.345	0.353	0.366	0.385	0.422	0.511	0.605	0.704	0.726	0.728	0.896
$t_1/b = 0.01$	0.300	0.301	0.303	0.314	0.327	0.334	0.345	0.353	0.366	0.385	0.422	0.510	0.605	0.704	0.726	0.728	0.896

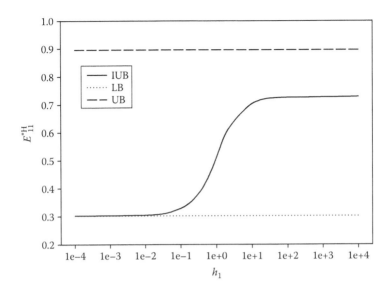

FIGURE 2.23
Flexural stiffness E_{11}^{H} with $t_1/b = 0.05$.

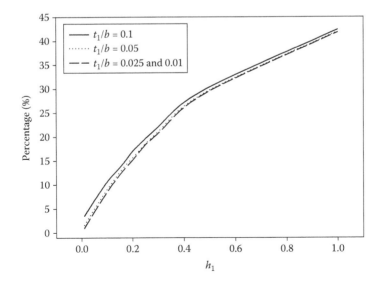

FIGURE 2.24
Underestimate percentage of E_{11}^{*H}.

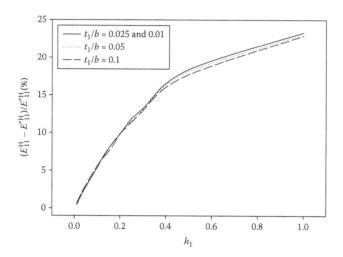

FIGURE 2.25
Percentage difference between stretch and flexural stiffness.

especially for three-point bending testing where shear stiffness is highly sensitive to the accuracy of flexural rigidity.

c. With similar derivation, another flexural stiffness E_{21}^{*H} is given below:

$$D_{21}^{H} = \delta^{3} \left\langle C_{2211}^{*} \right\rangle = \frac{\delta^{3}}{12} E_{21}^{*H}, \quad \text{where}$$

$$E_{21}^{*H} = 12 \times \left\{ 2 \int_{0}^{n/2} \int_{-1/2}^{1/2} \frac{E_{1}^{hc2}}{1-\Delta^{[2]}} \left[(v_{12}^{hc2} + v_{13}^{hc2} v_{32}^{hc2})(z^{2} + z \frac{1}{h_{1}} \frac{\partial V_{1}^{[2]}}{\partial y_{1}}) + z(v_{32}^{hc2} + v_{12}^{hc2} v_{31}^{hc2}) \frac{\partial V_{3}^{[2]}}{\partial z} \right] dz \, dy_{1} \right.$$

$$\left. + 2 \int_{n/2}^{1/2} \int_{-1/2}^{1/2} \frac{E_{1}^{hc1}}{1-\Delta^{[1]}} \left[(v_{12}^{hc1} + v_{13}^{hc1} v_{32}^{hc1})(z^{2} + z \frac{1}{h_{1}} \frac{\partial V_{1}^{[1]}}{\partial y_{1}}) + z(v_{32}^{hc1} + v_{12}^{hc1} v_{31}^{hc1}) \frac{\partial V_{3}^{[1]}}{\partial z} \right] dz \, dy_{1} \right\} \quad (2.151)$$

and the integration of which results in

$$E_{21}^{*H} = 12 \times \sum_{n} (-1)^{\frac{n}{2}} \frac{r_{1} f_{n}(v_{32}^{hc1} + v_{12}^{hc1} v_{31}^{hc1}) - [\frac{2d_{n}^{[1]}}{h_{1}n\pi} \sinh(\lambda_{n}^{[1]} \frac{h_{1}r_{1}}{2}) + \frac{2(-1)^{\frac{n}{2}} r_{1}}{n^{2}\pi^{2}}](v_{12}^{hc1} + v_{13}^{hc1} v_{32}^{hc1})}{1-\Delta^{[1]}} E_{1}^{hc1} \quad (2.152)$$

$$n = 2, 4, 6 \cdots$$

since $E_{2}^{hc2} = 0$.

For the case of E_{22}^{*H}, as seen from the GTM model in Figure 2.17, the only skin effect is due to Poisson's effect. Therefore, its value is the same as E_{22}^{H} per Equation (2.140), i.e.,

$$E_{22}^{*H} = E_{22}^{H} \quad (2.153)$$

2.3.5 Periodic Unit Cell Finite Element Analysis

2.3.5.1 Periodic Boundary Conditions

To verify the variational approach given in the preceding section, the method of unit cell FEA is used, which is much more effective than actual detailed modeling in terms of computations and the manifestation of size and edge effects. Besides the verification of the improved upper bound (IUB) of semi-analytical solutions in Section 2.3.4, the unit cell FEA can further demonstrate the effect of skin rigidity on the homogenized properties, by which Equations (2.161) and (2.162) and the correction coefficients, which are later shown in Table 2.24, are proposed.

The technique of unit cell modeling is on how to impose periodic boundary conditions in both displacement and stress fields, whereas most FEA homogenization problems have to be treated with specialized hybrid elements. In this study with common elements of a commercial FEA package (ANSYS5.5), a unit cell modeling technique is developed, due to the transverse symmetry of sandwiches.

First, the stress function Ψ is introduced by

$$\sigma_y = \frac{\partial^2 \Psi}{\partial z^2} \quad \sigma_z = \frac{\partial^2 \Psi}{\partial y^2} \quad \tau_{zy} = -\frac{\partial^2 \Psi}{\partial y \, \partial z} \tag{2.154}$$

In Figure 2.26, the displacement and stress are continuous with regard to the boundary AB and A'B', i.e.,

$$\sigma_1^h(-\frac{1}{2},z) = \sigma_1^h(\frac{1}{2},z) \quad \tau_{13}^h(-\frac{1}{2},z) = \tau_{13}^h(\frac{1}{2},z)$$

$$\Theta_1(-\frac{1}{2},z) = \Theta_1(\frac{1}{2},z) \quad \Theta_3(-\frac{1}{2},z) = \Theta_3(\frac{1}{2},z) \tag{2.155}$$

where Q represents the homogenization function Π, U, or V for the mode of shear, tension, and bending, respectively.

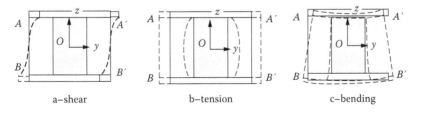

a–shear   b–tension   c–bending

FIGURE 2.26
Deformation of periodic unit cell.

2.3.5.1.1 Pure Transverse Shear Mode

In the case of transverse shear mode, there is clearly an antisymmetric relation between AB and A'B', i.e.,

$$\sigma_1^h(-\frac{1}{2},z) = \sigma_1^h(\frac{1}{2},-z) \qquad \tau_{13}^h(-\frac{1}{2},z) = \tau_{13}^h(\frac{1}{2},-z)$$

$$\Pi_1(-\frac{1}{2},z) = -\Pi_1(\frac{1}{2},-z) \qquad \Pi_3(-\frac{1}{2},z) = -\Pi_3(\frac{1}{2},-z)$$

(2.156)

It is found that

$$\frac{\partial^2 \Psi}{\partial z^2}(\pm\frac{1}{2},z), \ \frac{\partial^2 \Psi}{\partial y^2}(\pm\frac{1}{2},z) \text{ and } \frac{\partial^2 \Psi}{\partial y \partial z}(\pm\frac{1}{2},z)$$

are all even about z. The following thus can be deduced sequentially:

$$\sigma_1^h(\pm\frac{1}{2},z) = \sigma_3^h(\pm\frac{1}{2},z) = 0$$

$$\varepsilon_1^h(\pm\frac{1}{2},z) = \varepsilon_3^h(\pm\frac{1}{2},z) = 0$$

(2.157)

By (2.157) the periodic boundary condition for FEA modeling can be finally implicated in the form of

$$\Pi_3(\pm\frac{1}{2},z) = \Pi_3(0,z) = 0$$

(2.158)

2.3.5.1.2 Pure Tension and Bending Mode

With similar deduction, the periodic boundary conditions for tension and bending mode are, respectively, given as

$$U_1(\pm\frac{1}{2},z) = U_1(0,z) = 0$$

(2.159)

and

$$V_1(\pm\frac{1}{2},z) = V_1(0,z) = 0$$

(2.160)

The deformations of the three modes are illustrated by the dashed line in Figure 2.26.

2.3.5.2 Three-Dimensional Modeling and Results

With the deduction of 2D periodic boundary conditions in Section 2.3.5.1, actual hexagons can be modeled by extending 2D conditions to 3D ones. As shown in Figure 2.27, there are two periodic sections ABCD and A'B'C'D', and the displacement constraints are imposed as listed in Table 2.21. For all three modes, plane strain globally occurs in the x-y plane, i.e., no displacement in z for two lateral cell walls. In the transverse shear mode, the nodes at the lines DOD'/CO'C' have zero displacement in UY and UZ due to double symmetry, which is additionally imposed to ensure the accuracy of the modeling.

SHELL ELEMENT 93 in ANSYS 5.5 is used for the modeling of core walls and skin faces. The refinement study indicates that the convergence can be quickly achieved, and the final mesh is illustrated in Figure 2.27 for the case of $h_1 = 1$.

2.3.5.2.1 Nonlinear Effect

The nonlinear effect, mainly due to the membrane force of skin faces, is assessed. The assessment is conducted for three modes separately, though not the combination of them. It is observed that the nonlinear effect becomes more evident with the increased ratio

$$R_n = \frac{t_f E_f}{\delta E_c},$$

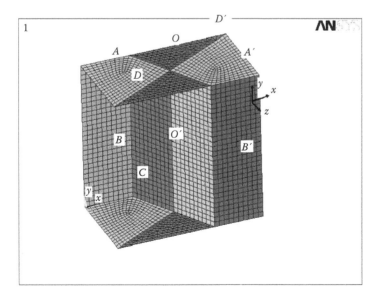

FIGURE 2.27
Unit cell FE modeling.

TABLE 2.21

Boundary Conditions of Periodic Unit Cell FE Modeling

Mode	Nodes	UX	UY	UZ	ROTX	ROTY	ROTZ	Remark
Transverse shear	A'B'CD/ABC'D'	F	F	0	0	0	F	Symmetric
	AD/A'D'	+Δ	0	0	F	0	0	
	BC/B'C'	−Δ	0	0	F	0	0	
	AB/OO'/A'B'	F	0	0	0	0	F	
	DOD'/CO'C'	F	0	0	F	F	F	Additional
In-plane tension	A'B'/ABC'D'	F	F	0	0	0	F	Symmetric
	AD/BC	−Δ	F	F	F	0	0	
	A'D'/B'C'	+Δ	F	F	F	0	0	
Cylindrical bending	A'B'/ABC'D'	F	F	0	0	0	F	Symmetric
	AD	+Δ	F	F	F	0	$-2\Delta/(t_f+d)$	
	BC	−Δ	F	F	F	0	$-2\Delta/(t_f+d)$	
	A'D'	−Δ	F	F	F	0	$+2\Delta/(t_f+d)$	
	B'C'	+Δ	F	F	F	0	$+2\Delta/(t_f+d)$	
	AB	L	F	0	0	0	$-2\Delta/(t_f+d)$	
	A'B'	L	F	0	0	0	$+2\Delta/(t_f+d)$	

Note: F = free; L = linear.

TABLE 2.22

Nonlinear Effect on G_{13}^H (Unit: G_{13}^{hc1})

Shear Strain		100 με	200 με	350 με	575 με	9,125 με	1,000 με
$R_n = 5/3$	Nonlinear	1.5844	1.5844	1.5845	1.5849	1.5845	1.5846
	Linear	1.5836	1.5836	1.5836	1.5836	1.5836	1.5836

and the nonlinear cases are modeled to assess the sensitivity of the ratio R_n. As shown in Table 2.22, for the case of $R_n = 5/3$ and shear strain up to 1000 me, the resulting difference of G_{13}^H is less than 0.08%. Since the ratio R_n of practical applications is mostly less than 0.1, the nonlinear effect can be overlooked particularly in the elastic range, and it is not considered in this study.

2.3.5.2.2 Transverse Shear Stiffness

It may be conjectured that the effect of skin rigidity can be expressed by the skin rigidity ratio

$$\left(\frac{t_f}{\delta}\right)^3 \frac{E_f}{\left|G_{13}^{hc1} - G_{13}^{hc1}\right|},$$

and an approximate equation (2.161) is thus developed by interpolating groups of FEA results for G_{13}^H of regular hexagons, which complements the IUB solution of (2.116)

$$G_{13}^H = G_{13}^H\Big|_{IUB} - \frac{h_1}{120} R^{-\frac{5}{6}}, \quad R \geq 0.1$$

$$R = 0.54 \left(\frac{t_f}{\delta}\right)^3 \frac{b}{t_1} \frac{E_f}{E_c} \tag{2.161}$$

where $G_{13}^H\Big|_{IUB}$ is calculated by (2.116) and listed in Table 2.18. For those hexagons $R < 0.1$, the G_{13}^H value can approximately take the lower bound value of Equation (2.119). Equation (2.161) is validated for those of $h_1 < 1$, the range of which is sufficient for general honeycomb sandwiches. Further note for anisotropic skin faces, E_f in (2.161) can be approximately replaced with E_{f1}. The closeness of (2.161) to FEA results is shown in Figure 2.28 and Table 2.23 with less than 1/750 of difference. The deformations of the unit cell shear are illustrated in Figure 2.29.

2.3.5.2.3 Tension and Flexural Stiffness

The formulas (2.131) and (2.147) developed in Section 2.3.4 for tension and flexural stiffness are based on the assumption of infinite rigid skin faces, which result in the improved upper bound (IUB). In this study, by using FEA, the effect of skin rigidity can be further assessed, and it is expressed

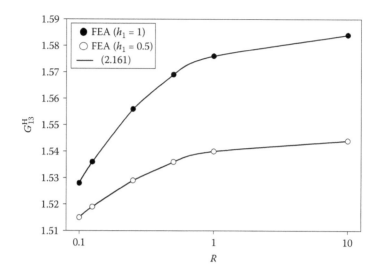

FIGURE 2.28
Effect of skin rigidity: Equation (2.161) vs. FEA.

TABLE 2.23

G_{13}^H: Equation (2.161) vs. FEA Results (Unit: G_{13}^{hc1})

R		0.1	0.125	0.25	0.5	1	10
$h_1 = 1$	Equation (2.161)	1.528	1.537	1.558	1.569	1.576	1.583
	FEA	1.528	1.536	1.556	1.569	1.576	1.584
$h_1 = 1/2$	Equation (2.161)	1.516	1.521	1.531	1.537	1.540	1.543
	FEA	1.515	1.519	1.529	1.536	1.540	1.544
$h_1 = 1/4$	Equation (2.161)	1.519	—	—	—	1.507	1.521
	FEA	1.519	—	—	—	1.506	1.521

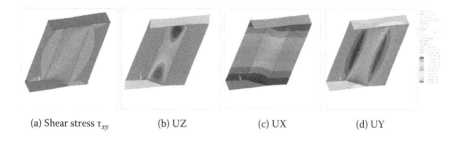

(a) Shear stress τ_{xy} (b) UZ   (c) UX   (d) UY

FIGURE 2.29
Pure transverse shear mode.

by two correction coefficients K and K* for tension and flexural stiffness, respectively.

$$E_{11}^H = K \ E_{11}^H \big|_{IUB}$$

$$E_{11}^{*H} = K^* \ E_{11}^{*H} \big|_{IUB}$$

(2.162)

The correction coefficients are listed in Table 2.24 with respect to the ratios of t_1/b, E_f/E_c, and t_f/δ that cover most common sandwiches. It is found that when $h_1 < 1/8$ and $1/4$, respectively, the correction coefficients K and K* can be approximated to be 1.00. For the higher value of h_1, the correction coefficients in Table 2.24 can be used in combination with (2.131) or (2.147) to include the rigidity of skin faces. The deformations of tension and bending are illustrated in Figures 2.30 and 2.31.

TABLE 2.24

Correction Coefficients K and K*

		t_1/b	E_f/E_c	t_f/d	K	K*
$h_1 = 1$	0.1		$1 \sim 2$	$1/12$	$0.86 \sim 0.90$	$0.93 \sim 0.94$
			$1 \sim 2$	$1/6$	$0.91 \sim 0.93$	
	Others		$1 \sim 2$	$1/60 \sim 1/24$	$0.86 \sim 0.91$	$0.90 \sim 0.93$
			$1 \sim 2$	$1/12 \sim 1/6$	$0.91 \sim 0.92$	$0.94 \sim 0.97$
$h_1 = 1/2$	0.1		1	$1/24 \sim 1/12$	$0.88 \sim 0.91$	$0.94 \sim 0.96$
			2	$1/12 \sim 1/6$	$0.93 \sim 0.95$	$0.97 \sim 0.99$
	Others		$1 \sim 2$	$1/24 \sim 1/6$	$0.91 \sim 0.92$	$0.97 \sim 0.99$
$h_1 = 1/4$	0.1		$1 \sim 2$	$1/48 \sim 1/24$	$0.89 \sim 0.91$	$0.97 \sim 1.00$
				$1/12 \sim 1/6$	$0.92 \sim 0.94$	
	Others		1	$1/120 \sim 1/24$	$0.89 \sim 0.91$	$0.97 \sim 0.99$
			2	$1/24 \sim 1/6$	$0.91 \sim 0.93$	$0.99 \sim 1.00$
$h_1 = 1/5$	0.1		$1 \sim 2$	$1/60 \sim 1/6$	$0.91 \sim 0.96$	
	Others		$1 \sim 2$	$1/120 \sim 1/12$		≈ 1.00
$h_1 = 1/6$	0.1		$1 \sim 2$	$1/72 \sim 1/6$	$0.93 \sim 0.98$	
	Others		$1 \sim 2$	$1/120 \sim 1/12$		
$h_1 = 1/8$	0.1 and others		$1 \sim 2$	$1/120 \sim 1/12$	$0.97 \sim 1.00$	
$h_1 < 1/8$	0.1 and others		$1 \sim 2$	$1/120 \sim 1/12$	≈ 1.00	

Note: Others represents $t_1/b = 0.05$, 0.025, and 0.01.

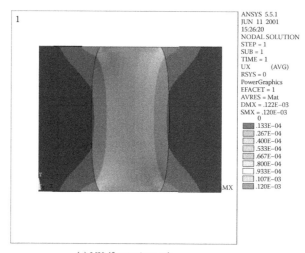

(a) UX (front view +z)

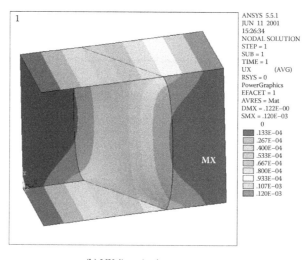

(b) UX (iso-view)

FIGURE 2.30
Pure tension mode.

2.3.6 Summary and Concluding Remarks

In this section, constitutive modeling of honeycomb sandwiches is developed, and an effective theoretical approach is proposed to derive elastic stiffness tensors for general honeycomb sandwiches. The usually neglected skin effect is given particular attention, and is found to play an important role in both sandwich local fields and global behaviors.

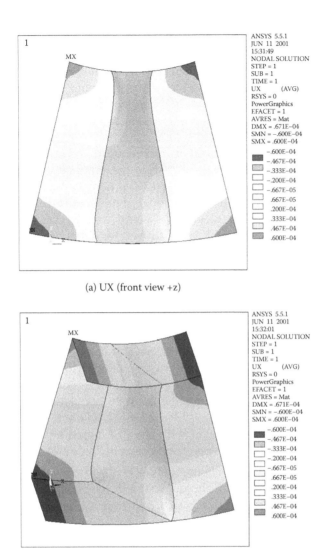

(a) UX (front view +z)

(b) UX (iso-view)

FIGURE 2.31
Pure bending mode.

The adaptation of the homogenization theory to periodic plates is introduced and extended to include transverse shear deformation theory, by which the field equations of three local problems are deduced. Then, a multipass homogenization technique is applied to solve the 3D homogenization functions. With the first pass of the geometry-to-material transformation model (GTM), the spatial heterogeneity is conveniently transformed into the material anisotropy. And in the second pass, the unit cell is 2D homogenized by global plane strain, where the solution is analytically sought in a variational sense. The stiffness tensors are finally formulated in the form of Fourier series that is easily calculated with a symbolic program. Finally, the finite element analysis is conducted to verify and complement the analytical solutions with the additional assessment on the effect of skin rigidity.

In this study, there are several observations, as follows:

1. The multipass homogenization technique is successfully applied in periodic cellular structures, while the engineering application of the technique is sought for structural homogenization. Further, with the geometry-to-material model, the 3D anisotropic elasticity is practiced on real engineering problems, among very few cases in the literature. Honeycomb constitutive modeling, after incorporating these concepts, thus has much more flexibility in the improvement of computational accuracy and expense. It should be noted that the approach introduced in this section is readily applicable for all general 2D cellular configurations. The adaptation of the homogenization theory in periodic plates is introduced and modified to include the transverse shear deformable plate theory. And the homogenization function of transverse shear stiffness is, for the first time, derived with an analytical solution, which, besides other stiffness, is validated with FEA results and literature. It is advised that for general computations the accurate calculations of stiffness follow the formulas (2.116), (2.131), (2.138), (2.147), and (2.152) derived in this study complemented with the FEA-based equations (2.161) and (2.162) and the correction coefficients in Table 2.24.

2. The refiner analysis may choose unit cell FEA with the specialized code or periodicity modeling technique that is developed in this section.

3. The flexural stiffness and stretch stiffness are distinguished for the honeycomb sandwich structures, the mechanism of which is demonstrated through their symmetric and antisymmetric nature. Although the impact of this divergence on sandwich computational modeling needs further evaluations, the careful use of equivalent core properties is hereby emphasized, which include the sensitive Poisson's ratios.

4. The skin effect plays an important role in all sandwich stiffnesses for general practical honeycomb sandwiches, particularly when the thickness ratio e/d is not so small. Since in sandwich computations the skin effect has not been taken into account much, there is a strong recommendation to include this effect in refined analysis and to have further investigations on current applications. The particular one is of sandwich beams, on which many tests and investigations are based.

5. Nonlinear effect is briefly assessed in this study, and it is found negligible in the elastic range. However, a thorough study is required for more understanding on nonlinear behavior and coupling among shear, stretch, and bending modes. Also, the assessment of size and edge effects in cellular modeling can be a relevant topic, especially for the justification of divergence between theoretic and experimental results.

References

Adams, R.D., and Maheri, M.R. (1993). The dynamic shear properties of structural honeycomb materials. *Composite Science and Technology*, 47, 15–23.

Allen, H.G. (1969). *Analysis and design of sandwich panels*. Pergamon Press, Oxford.

Astley, R.J., Harrington, J.J., and Stol, K.A. (1997). Mechanical modeling of wood microstructure, an engineering approach. *IPENZ Transactions EMCH*, 24(1), 21–29.

Becker, W. (1998). The inplane stiffness of a honeycomb core including the thickness effect. *Archives of Applied Mechanics*, 68, 334–341.

Bourgeois, S., Cartraud, P., and Debordes, O. (1998). *Homogenization of periodic sandwiches. Mechanics of sandwich structures*. Kluwer Academic Publishers, New York, pp. 139–146.

Caillerie, D. (1984). Thin elastic and periodic plates. *Mathematical Methods in the Applied Sciences*, 6, 159–191.

Chamis, C.C. (1984). *Simplified composites micromechanics equations for strength, fracture toughness, and environmental effects*. NASA TM-83696.

Chamis, C.C., Aiello, R.A., and Murthy, L.N. (1988). Fiber composite sandwich thermostructural behavior: Computational simulation, *ASME Journal of Composites Technology and Research*, 10(3), 93–99.

Christensen, R.M. (1991). *Mechanics of composite materials*. Krieger, Malabar.

Cunningham, P.R., and White, R.G. (2001). A new measurement technique for the estimation of core shear strain in closed sandwich structures. *Composite Structures*, 51, 319–334.

Daniel, M.I., and Abot, L.J. (2000). Fabrication, testing and analysis of composite sandwich beams. *Composites Science and Technology*, 60, 2455–2463.

Davalos, J.F., Qiao, P.Z, Xu, X.F, Robinson, J., and Barth, K.E. (2001). Modeling and characterization of fiber-reinforced plastic honeycomb sandwich panels for highway bridge applications. *Composite Structures*, 52(3–4), 441–452.

Davalos, J.F., Salim, H.A., Qiao, P.Z., Lopez-Anido, R., and Barbero, E.J. (1996). Analysis and design of pultruded FRP shapes under bending. *Composites: Part B, Engineering Journal*, 27(3–4), 295–305.

Duvaut, G. (1977). Comportement macroscopique d'une plaque oerforee periodiquement. In *Lecture notes in math*. Vol. 594. Springer, Berlin, pp. 131–145.

Fortes, M.A., and Ashby, M.F. (1999). The effect of non-uniformity on the in-plane modulus of honeycombs. *Acta Materials*, 47(12), 3469–3473.

Germain, P., Nguyen, Q.S., and Suquet, P. (1983). Continuum thermodynamics. *Journal of Applied Mechanics*, 50, 1010–1020.

Gibson, L., and Ashby, M. (1988). *Cellular solids structure and properties*. Pergamon Press.

Grediac, M. (1993). A finite element study of the transverse shear in honeycomb cores. *International Journal of Solids Structures*, 30(13), 1777–1788.

Harris, J.S., and Barbero, E.J. (1998). Prediction of creep properties of laminated composites from matrix creep data. *Journal of Reinforced Plastics and Composites*, 17(4), 361–379.

Hashin, Z., and Rosen, B.W. (1964). The elastic modeling of fiber-reinforced materials. *Journal of Applied Mechanics*, 31, 223–232.

Hohe, J., and Becker, W. (2001). A refined analysis of the effective elasticity tensors for general cellular sandwich cores. *International Journal of Solids Structures*, 38, 3689–3717.

Hohe, J., Beschorner, C., and Becker, W. (1999). Effective elastic properties of hexagonal and quadrilateral grid structures. *Composite Structures*, 46, 73–89.

Jones, R.M. (1999). *Mechanics of composite materials*. Taylor & Francis, Philadelphia, PA.

Kalamkarov, A. L. (1992). *Composite and reinforced elements of construction*. Wiley, New York.

Kelsey, S., Gellatly, R.A., and Clark, B.W. (1958). The shear modulus of foil honeycomb cores. *Aircraft Engineering*, 30, 294–302.

Lekhnitskii, S.G. (1968). *Anisotropic plates*. Gordon and Breach Science Publishers.

Lewinski, T. (1991). Effective models of composite periodic plates. I. Asymptotic solution. II. Simplifications due to symmetries. III. Two-dimensional approaches. *International Journal of Solids Structures*, 27, 1155–1203.

Lewinski, T. (2000). *Plates, laminates and shells asymptotic analysis and homogenization*. World Scientific, River Edge, NJ.

Luciano, R., and Barbero, E.J. (1994). Formulas for the stiffness of composites with periodic microstructure. *International Journal of Solids and Structures*, 31(21), 2933–2944.

Manet, V., Han, W.-S., and Vautrin, A. (1998). Static analysis of sandwich plates by finite elements. In *Mechanics of sandwich structures*. Kluwer Academic Publishers, New York, pp. 139–146.

Masters, I.G., and Evans, K.E. (1996). Models for the elastic deformation of honeycombs. *Composite Structures*, 35, 403–422.

Meguid, S.A., and Kalamkarov, A.L. (1994). Asymptotic homogenization of elastic composite materials with a regular structure. *International Journal of Solids Structures*, 31(3), 303–316.

Noor, A., Burton, W.S., and Bert, C.W. (1996). Computational models for sandwich panels and shells. *Applied Mechanics Review*, 49(3), 155–199.

Parton, V.Z., and Kudryavtsev, B.A. (1993). *Engineering mechanics of composite structures*. CRC Press, Boca Raton, FL.

Penzien, J., and Didriksson, T. (1964). Effective shear modulus of honeycomb cellular structure. *AIAA Journal*, 2(3), 531–535.

Plunkett, J.D. (1997). *Fiber-reinforcement polymer honeycomb short span bridge for rapid installation*. IDEA Project Report.

Qiao, P.Z., Fan, W., Davalos, J.F., and Zou, G.P. (2008). Homogenization and optimization of sinusoidal honeycomb cores for transverse shear stiffness. *Journal of Sandwich Materials and Structures*, 10(5), 385–412.

Qiao, P.Z., Fan, W., Davalos, J.F., and Zou, G.P. (2008). Optimization of transverse shear moduli for composite honeycomb cores. *Composite Structures*, 85(3), 265–274.

Qiao, P.Z., and Xu, X.F. (2005). Refined analysis of torsion and in-plane shear of honeycomb sandwich structures. *Journal of Sandwich Structures and Materials*, 7(4), 289–305.

Shi, G., and Tong, P. (1995a). Equivalent transverse shear stiffness of honeycomb cores. *International Journal of Solids Structures*, 32(10), 1383–1393.

Shi, G., and Tong, P. (1995b). The derivation of equivalent constitutive equations of honeycomb structures by a two scale method. *Computational Mechanics*, 15, 395–407.

Takano, N., Zako, M., and Kikuchi, N. (1995). Stress analysis of sandwich plate by the homogenization method. *Materials Science Research International*, 1(2), 82–88.

Vonach, W.K., and Rammerstorfer, F.G. (1998). *Effects of in-plane stiffness on the wrinkling load of sandwich constructions*. Sandwich Constructions IV, Vol. 1. EMAK UK.

Vougiouka, G., and Guedes, H.R. (1998). *Prediction of elastic properties of sandwich panels using a homogenization computational model*. Mechanics of Sandwich Structures. Kluwer Academic Publishers, New York, pp. 147–154.

Warren, W.E., and Kraynik, A.M. (1987). Foam mechanics: The linear elastic response of two-dimensional spatially periodic cellular materials. *Mechanics of Materials*, 6, 27–37.

Xu, X.F., and Qiao, P.Z. (2002). Homogenized elastic properties of honeycomb sandwich with skin effect. *International Journal of Solids and Structures*, 39(8), 2153–2188.

Xu, X.F., Qiao, P.Z., and Davalos, J.F. (2001). On the transverse shear stiffness of composite honeycomb core with general configuration. *ASCE Journal of Engineering Mechanics*, 127(11), 1144–1151.

3

FRP Deck: Strength Evaluation

3.1 Overview

This chapter is focused on strength evaluation of honeycomb fiber-reinforced polymer (HFPR) sandwich panels with sinusoidal core geometry, in terms of both experimental investigation and theoretical analysis. The sandwich structures consist of core and facesheet components, where core materials are primarily subjected to out-of-plane compression and shear, and the facesheet carries mainly membrane forces. Therefore, the first objective is to study the core material under out-of-plane compression. Chopped strand mat (ChSM) is used for the core material, which is composed of E-glass fibers and polyester resin. The facesheet is made of several layers of ChSM, 0°/90° E-glass fiber, and polyester resin. The ChSM material is used at the interface between the core and facesheet as a bonding layer. These component parts are joined by the contact molding manufacturing process. As a result, the number of the ChSM bonding layers and core thickness plays an important role on the compressive strength of the panel. Analytical models are provided to predict the pure compressive and buckling strength, which are verified through finite element (FE) results. Compression tests are further carried out to correlate with the analytical results. The number of the ChSM bonding layers and panel core thickness define each specimen type. Different failure modes are obtained for different parameter combinations, and their linear and failure responses are described.

The second objective is to study the core materials under out-of-plane shear. Analytical models are provided for shear buckling, pure shear failure, and debonding. Shear buckling can be solved using the Rayleigh–Ritz method, and the other two failure modes are characterized based on accurate description of shear stiffness and stress distribution considering skin effect. Design formulas are provided to predict the failure strength. To verify the analytical models, a series of four-point bending tests are further carried out by varying the number of bonding layers and core thickness, both along longitudinal and transverse directions.

The final objective is to study the strength properties of the facesheet. A finite element (FE) progressive failure model is developed, which is validated by the existing test results, and is further used to carry out a parametric study by varying material properties, layer thickness, and layer sequences. Compression and bending tests are carried out on selected layer configurations, and an optimized facesheet configuration is recommended.

3.2 Literature Review

3.2.1 Introduction

As stated in Section 3.1, the objective of this study is to evaluate the strength of core materials, facesheets, and the interface between core and facesheet. Much effort has been devoted to the stiffness modeling and optimization of the honeycomb fiber-reinforced polymer (HFRP) sandwich panel. In Chapter 2, equivalent orthotropic properties representative of the complex honeycomb geometry are developed, a simplified analysis procedure that can be used in design applications is presented, and an analytical solution for the transverse shear stiffness of a composite honeycomb with general configurations is derived. However, only a few studies are available on the strength properties of sandwich structures, and particularly HFRP, partly due to their complicated honeycomb core geometry. A previous study by DeTeresa et al. (1999) indicated that core materials for sandwich structures are primarily subjected to shear and through-the-thickness compression. Thus, this section reviews failure mechanisms under such loading conditions to evaluate the behavior of HFRP sandwich panels. To this end, Section 3.2.2 reviews the previous work on core materials under out-of-plane compression, and Section 3.2.3 is focused on out-of-plane (transverse) shear. Section 3.2.4 examines significant issues related to facehsheet studies.

3.2.2 Out-of-Plane Compression

3.2.2.1 Compressive Strength of Core Materials

Chopped strand mat (ChSM) is used for the core materials that are composed of E-glass fiber and polyester resin. The modeling of ChSM can be dated back to the 1970s. Hann (1975) defined the random composite by an equivalent laminate consisting of unidirectional plies in the plane of the laminate. Using a maximum stress criterion, the strength of the random composite was given in terms of uniaxial strength of the unidirectional composite through a simple relation. Halpin and Kardos (1978) modeled the random fiber composites as a quasi-isotropic laminate consisting of $(0°/90°/\pm45°)_s$ plies. A maximum strain failure criterion was considered to predict the ultimate strength. They provided several examples illustrating the use of the model. Both of these studies treated the ChSM as lay-ups of laminate in a balanced condition, which are still used currently (Barbero 1999).

3.2.2.2 Core Crushing

One of the common failure modes for sandwich structures under out-of-plane compression is core crushing. Theotokoglou (1996) offered an analytical determination of the ultimate strength of sandwich beams considering the core failure in compression, tension, and shear using the maximum failure strength method. He also performed a pull-out test to verify his model. However, his study only gave an indication of the failure modes that took place in a T-joint under pull-out load, and further research was required in order to predict accurately the failure modes. Cvitkovich and Jackson (1998) studied the compression failure mechanisms in composite sandwich structures. The specimens in their study were tested with no damage, with a 6.35 mm diameter hole and three levels of impact damage. Mouritz and Thomson (1999) investigated the compression, flexure, and shear properties of a sandwich composite containing defects. They concluded that determining the compressive properties of a large sandwich structure was difficult because the strength and failure mechanism were dependent on the gage length. Core crushing under compression was observed in all these studies.

3.2.2.3 Buckling

For HFRP sandwich panels used for bridge deck applications, the following distinct features characterize them from their counterparts in other fields; they have relatively larger and sparsely distributed honeycomb cells, and the core and facesheets are manufactured separately and subsequently connected by contact bonding, using a ChSM and polymer resin at the interface. Due to the relatively low material stiffness and thin-walled sectional geometries of structural components, two possible instability problems for sandwich panels may result under different compression loading conditions. One is the wrinkling of the facesheet under in-plane compression (Niu and Talreja 1998), and the other is the instability of the core due to out-of-plane compression (Zhang and Ashby 1992). Out-of-plane compression is unavoidable in civil engineering applications, such as local compression on bridge decks exerted by wheel loads. The buckling of the honeycomb core becomes more significant due to the sparsely distributed thin-wall core panels. As reported by Kumar et al. (2003), local buckling of the thin walls precipitated most failure modes in their bending tests of tube bridge decks. Thus, it is beneficial to have a solution for transverse buckling of core elements, with loaded edges partially restrained by the interface bond with the facesheet panels.

Zhang and Ashby (1992) concluded that two possible failure modes for out-of-plane compression were buckling and material crushing, referred to as pure compression failure. In their study to predict buckling strength, they assumed the two edges of the core wall perpendicular to the loading direction as simply supported, while the other two loaded edges were rigidly constrained. Their solution was later applied by Lee et al. (2002) to study the behavior of the honeycomb composite core at elevated temperature. Both of these studies assumed a completely rigid connection at the facesheet-core interface, which is seldom the case in practice. The partial constraint offered by the interface bond has a significant effect on the behavior of FRP sandwich panels. This effect may vary due to different materials and manufacturing techniques used, with the clamped and hinged conditions as two extreme cases for the connection. In general, the quality of the bonding effect can be improved by either selecting compatible bonding materials or increasing embedment of the core into the bonding layer. The latter method is analogous to increasing the contact depth, or increasing the bonding layer thickness, which in turn produces larger fillets of excess adhesive, which formed at honeycomb interfaces, effectively increasing the bonding area. This facesheet-core interaction is typically called the bonding layer effect. Burton and Noor (1997) used detailed FE models to examine the effect of the adhesive joint on the load transfer and static responses of sandwich panels. However, they used strain energy for discrete components to discuss the effect of various parameters, a method that cannot be readily used in practice. Up to now, the bonding effect on the behavior of honeycomb sandwich panels has not yet been clearly defined. It is the objective of this study to quantitatively study this effect on the behavior of sandwich panels under compressive load.

By considering the bonding layer effect, the problem can be interpreted as the instability of a partially restrained plate. The research on this topic can be traced back to the 1950s (Bleich 1952). Qiao et al. (2001) performed a study on the local buckling of composite FRP shapes by discrete plate analysis. They provided an explicit solution for the problem with elastic constraint along the unloaded edges, and also provided detailed references on this topic. Their research was further explored by Kollar (2002) and Qiao and Zou (2003). All of the previous studies are focused on the buckling of plates or panels with elastic restraint along the unloaded edges; this restraint is provided by connection of flange-to-web elements for beam-type members. However, for the HFRP core under out-of-plane compression, the two edges in contact with the facesheet panels (Figure 1.1) can be treated as partially constrained; i.e., the elastic restraint is along the loaded edges. And this restraint results from the degree of connectivity between the facesheet and core. At present, there are only limited studies (Shan and Qiao 2008) for this problem, and it is therefore advantageous to develop an analytical solution for compression buckling capacity of a plate with two loaded edges partially constrained.

In this chapter, analytical models are provided for the two failure modes: core crushing and buckling. The coefficient of elastic restraint is introduced to quantify the bonding layer effect. A comprehensive approach is developed to study the buckling behavior of the HFPR core with varying degrees of boundary restraints, and an analytical solution is proposed by solving a transcendental equation. The bonding layer effect is evaluated experimentally by compression tests, which are designed such that buckling failure and pure compression failure can occur distinctly and separately. A novel testing method to predict the elastic constraint coefficient is also described; a parametric study is carried out to study the aspect ratio effect on the buckling behavior, and finally, design guidelines are proposed.

3.2.3 Out-of-Plane Shear

It is commonly believed that two failure modes may occur for a sandwich panel under out-of-plane shear: shear crushing (Allen 1969; Vinson 1999) and shear buckling (Qiao et al. 2001; Papadopoulos and Kassapoglou 2004; Qiao and Huo 2011). Chen and Davalos (2005) pointed out that the skin effect can significantly affect interfacial stress distribution, yielding a coupled stress state, where the normal stress may even be larger than the shear stress. They concluded that, unlike the common belief that only shear stress occurs when the structure is under pure shear force, tensile force at the interface arises for a sandwich core, especially at the intersections of core elements, making such locations critical for debonding. Therefore, debonding may occur well before shear crushing or buckling is achieved.

  To predict the shear strength of a sandwich panel, an accurate description of the stiffness is required. The computational models on honeycomb sandwiches are generally based on the equivalent replacement of each component with homogeneous continuum, due to expensive computation of 3D detailed properties. Therefore, to accurately represent the equivalent properties has been a perennially challenging topic that has attracted a lot of investigation. From Figure 1.1, one can intuitively conclude that honeycomb sandwich structures behave like I-beams: the outer facesheets correspond to the flanges, and carry most of the direct compression/tension bending load, and the lightweight core corresponds to the I-beam web. The core supports the skins, increases bending and torsional stiffness, and carries most of the shear load (Noor et al. 1996). This characteristic of a three-layer arrangement leads to the classical sandwich theory (Allen 1969; Zenkert 1995). Unlike the facesheet, which can even be a laminated plate, the equivalent properties of honeycomb cores are more complicated, as illustrated and discussed in Chapter 2.

3.2.3.1 Stiffness Study on Equivalent Properties of Honeycomb Cores

A comprehensive review of the computational models on honeycomb sandwiches was given by Noor et al. (1996), where numerous references were cited. Basically, all existing studies can be organized into two major groups, with each either neglecting or including skin effects, as explained in Chapter 2.

3.2.3.2 Interfacial Stress Distribution

Several studies (e.g., Chen and Davalos 2003) have shown that delamination of the core from the facesheet is a typical failure mode for sandwich panels. The fracture mechanics method is usually applied to study this problem, including studies by Ungsuwarungsri and Knauss (1987), Cui and Wisnom (1992), El-Sayed and Sridharan (2002), Blackman et al. (2003), Wang (2004), and other numerous works. It is shown by all these previous works that a crack is initiated when the interface traction attains the interfacial strength, and the crack is advanced when the work of traction is equal to the material's resistance to crack propagation. Therefore, stress concentration at the interface can act as a criterion to predict the onset of the delamination, and there is a need to further investigate the stress field at the interface. Chen and Davalos (2005) presented an analytical model allowing the calculation of the stiffness of honeycomb cores as well as the interfacial stress distribution considering skin effect, both under in-plane and out-of-plane forces, for hexagonal cores. To the authors' knowledge, accurate description of stiffness and interfacial stress distribution remains an open topic for HFRP sandwich panels with a sinusoidal core. It is noted that the hexagonal cores (Figure 3.1) are different from the sinusoidal cores (Figure 1.1) in that, for the hexagonal core, both the straight and inclined panels are affected by the skin effect. However, for the sinusoidal core, due to the existence of the flat panel, only the sinusoidal panel is affected. The two geometries of closed-cell (hexagonal) and open-cell (sinusoidal) configurations represent two major types that can be considered for sandwich cores. Thus, as a further contribution to this

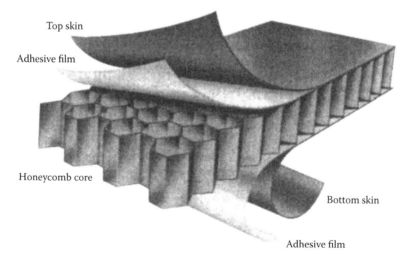

FIGURE 3.1
Sandwich panel with hexagonal honeycomb core. (From Noor, A. K., Burton, W. S., and Bert, C. W., *Applied Mechanics Reviews, ASME*, 49(3), 155–199, 1996.)

field, the behavior of a honeycomb sandwich panel with a sinusoidal core geometry considering skin effect is presented in this chapter.

3.2.3.3 Shear Crushing and Shear Buckling

The concept of shear failure consisting of shear crushing or shear buckling is relatively straightforward. Allen and Feng (1998) defined three categories of sandwich panels: (1) composite beam theory (CBT), where the sandwich is treated as an ordinary composite beam and there is no shear deformation; (2) elementary sandwich theory (EST), where stresses and deflections are calculated by composite beam theory, but there is an additional shear deflection associated with shear strains in the core; and (3) advanced sandwich theory (AST), where the faces must bend locally in order to follow the shear deformation of the core. Most of the sandwich panels, including HFRP sandwich panels in this study, fall into the category of EST. One basic assumption used for EST is that the core resists the shear force and the facesheet carries the membrane forces caused by the bending moment (Allen 1969; Vinson 1999). It is shown (Caprino and Langella 2000) that if the Young's modulus of the core is negligible with respect to the facing elastic modulus, and the facing thickness is small compared to the height of the core, the shear stress field in the core is practically uniform. Therefore, it is reasonable to assume that once this uniform shear stress exceeds the material shear strength, the panel will fail due to shear crushing.

The research on the shear buckling problem has a relatively long history. Bleich (1952) first studied the shear buckling strength of metal structures. Timoshenko and Gere (1961) refined this theory and studied buckling of rectangular plates under action of shear stresses. Barbero and Raftoyiannis (1993) used the first variation of the total potential energy equation to study the shear buckling of FRP structures. Qiao et al. (2001) further applied this theory to study the local buckling of webs under shear loading. More recently, Papadopoulos and Kassapoglou (2004) developed a method based on a polynomial expansion of the out-of-plane displacement of the plate and energy minimization and studied the shear buckling of rectangular composite plates with two concentric lay-ups. Most recently, Qiao and Huo (2011) developed the explicit closed-form local buckling solution of in-plane shear-loaded orthotropic plates with two opposite edges simply supported and the two opposite edges either both rotationally restrained or one rotationally restrained and the other free. In all these studies, energy method was employed, and therefore it is also adopted in this study. As pointed out earlier, two edges of the core panel are partially restrained. As a result, the potential energy will be given based on this boundary condition, and the Rayleigh–Ritz method will be used to solve this problem.

3.2.3.4 Testing Method

To study the shear behavior of the sandwich core, the American Society for Testing and Materials (ASTM) (ASTM C273-00) specifies a testing method.

However, this method cannot be directly applied to this study since the core is very strong in shear. Trial tests using this method illustrated that the failure is intralaminar delamination, instead of pure shear failure of the core. Another method, four-point bending test, is also recommended by ASTM (ASTM C393-00) to study shear strength and shear stiffness of sandwich cores since pure shear and bending regions will result from this loading condition. This method is adopted in this study for the HFRP sandwich. Many researchers have performed bending tests on sandwich beams. Lingaiah and Suryanarayana (1991) carried out experimental vs. analytical correlation of the mechanical properties of a sandwich beam specimen. Four-point and three-point load tests were conducted. It was observed that generally the failure load was higher for the case of the four-point bending test than for the three-point bending test. The failure of most specimens was due to debonding between the core and the facing and at loads that were less than the theoretical estimated based on the allowable core shear stress or the allowable facing tensile/compressive stress, whichever was lower, depending on the test condition. But they did not specify the position where the debonding initiated and did not provide an in-depth discussion of the mechanism behind the observed failure mode, where skin effect produces a tensile force in the pure shear region, causing the facesheet to debond from the core before the facesheet reaches its material strength. Mouritz and Thomson (1999) carried out four-point bending tests to study shear properties of a sandwich composite containing interfacial cracks subjected to impact load. They found that the composite containing the interfacial crack failed at a lower load than the defect-free specimen. The former failed due to a shear crack initiated near the interfacial crack tip, and upon loading grew into the foam until it reached the opposing skin, while skin wrinkling was a common failure mode in defect-free sandwich samples. The defect-free sandwich composite did not fail by a shear- or bending-dominated process. The stiffness and strength of the sandwich composite decreased with increasing impact energy and impact damage area except when the composite was loaded in bending tension. However, their tests were based on small coupon samples and did not translate well to predict the properties of large structural components. Zenkert (1991) also observed the same type of shear failure in polymer foam sandwich composites containing interfacial cracks. Zenkert (1991), Triantofillou and Gibson (1989), and Thomson et al. (1989) showed that the load needed to cause the onset of shear cracking can be predicted with good accuracy using analytical FE models based on the mode II fracture mechanics theory applied to layered anisotropic materials. Caprino and Langella (2000) performed three-point bending tests on a sandwich beam for the shear characterization of foam core. The special feature of their specimen was that they inserted rigid blocks in proximity to the concentrated load. They concluded that this method allowed for an accurate measurement of the shear modulus and shear strength compared to ASTM standards. However, this method was very complicated, and a lot of parameters needed to be calculated to design the test setup. Further test methods need to be generated to assess their applicability to core materials different from foam core.

In this chapter, analytical models are presented to predict the strength due to pure shear crushing, shear buckling, and the delamination of the core from facesheet. The skin effect can be described by considering both shear and bending warping effects. All previous studies on skin effect only considered the membrane force, which corresponds to shear warping defined herein. The bending warping effect is for the first time presented. The analytical models are verified through FE analysis. To further understand the behavior of core material under out-of-plane shear, four-point bending tests are carried out.

3.2.4 Facesheet Study

3.2.4.1 Progressive Failure Analysis

A lot of research has been done in the area of progressive failure analysis. The conventional strength analysis, called total ply discount (Vinson and Sierakowski 1987), does not recognize that ply failure is localized, and therefore it underestimates laminate strength. First-ply failure (FPF) can be used to predict the onset of the damage (Barbero 1999) as long as the stresses in each laminate are computed accurately. The objective of progressive failure analysis is focused on post-FPF analysis. According to Kim (1995), there are two approaches to include damage: (1) modifying the stiffness matrix directly (Lee 1982; Ochoa and Engblom 1987; Hwang and Sun 1989; Tolson and Zabaras 1991) and (2) degrading the material properties (Tan 1991; Tan and Perez 1993; Reddy and Reddy 1993; and Kim et al. 1996).

Using the stiffness modification approach, Lee (1982) developed a three-dimensional FE computer program to analyze a fiber-reinforced composite laminate. The program could calculate the detailed stress distribution, identify the damage zone and failure mode, analyze the damage accumulation, and determine the ultimate strength. He defined three types of damage: breakage of fibers, failure of matrix, and delamination. The stresses at the center of each element were taken as the representative of that element for fiber breakage and matrix failure, and the stresses at the center of the interface between two layers were taken as the representative stress for delamination. Based on the three damage types, the stiffness matrix was modified accordingly. He applied this program to study damage accumulation in composite laminate containing circular holes subjected to in-plane loading. However, due to mesh coarseness at the edge of holes, delamination could not be captured. Further refinement of the finite element mesh was practically impossible due to computational limitations. Ochoa and Engblom (1987) used a higher-order plate element and computed transverse stresses from equilibrium equations. The failure analysis procedure was similar to that used by Lee (1982). Hwang and Sun (1989) developed an iterative 3D finite element analysis with a modified Newton-Raphson scheme for the failure prediction of laminates. Tolson and Zabaras

(1991) followed a procedure similar to that used by Ochoa and Engblom (1987), using a higher-order plate element. Tsau and Plunkett (1993) investigated a square plate made of a layered composite material, with a centered circular hole subjected to in-plane biaxial loading, using a family of eight-node elements. Hashin failure criteria were adopted in their study and mesh size of the FE model in laminates was carefully considered. In their analysis, at each increment of load, only one element, which was the one with the largest function value of the criterion in either fiber or matrix mode, was assigned to fail.

Using the material degradation approach, Tan (1991) investigated the progressive failure with cut-out holes under in-plane tension testing. Different degradation factors were used for a longitudinal modulus due to fiber breakage and transverse shear moduli due to matrix failure. The same approach was adopted by Tan and Perez (1993) to study the compressive loading case. Reddy and Reddy (1993) developed a three-dimensional progressive failure algorithm where the layer-wise laminate theory of Reddy was used for kinematic description. The stiffness of reduction was carried out at the reduced integration gauss points of the FE mesh depending on the mode of failure. Two types of stiffness reduction methods were used: independent, i.e., each stress would contribute only toward degradation of the corresponding stiffness property; and interactive method, i.e., coupling was assumed between normal and shear stiffness properties. However, material properties were degraded by the same factor regardless of failure modes. They concluded that further investigation was required to apply their approach to laminates under compressive and bending load. Kim et al. (1996) formulated a beam element with layer-wise constant shear (BLCS) based on layer-wise laminated beam theory. Two schemes to predict load-displacement paths were used: load controlled and displacement controlled. The stiffness degradation factors were evaluated through parametric studies and correlation with experimental results. The BLCS predictions for ultimate loads and displacements were accurate compared to experimental results. However, when experimental responses showed nonlinear load-displacement behavior, the prediction for displacement could not exactly match experimental results.

Most of the previous progressive failure analysis using FE is based on in-house programs, which requires a lot of effort and time, and also the code developed by one researcher cannot be readily used by others. Nowadays, some general purpose FE analysis tools, such as ABAQUS, ANSYS, etc., are widely used in the academic and industrial fields. These programs allow users to define their own subroutines in the analysis to fulfill the functions, such as stiffness reduction and material degradation, as described above. After evaluating all the possibilities, we choose to develop a progressive failure model through a user-defined subroutine using ABAQUS.

One important issue in the progressive failure analysis is to find an appropriate failure criterion. Various failure criteria for isotropic or composite materials have been proposed. In general, the failure criteria are categorized into two groups: independent and polynomial failure criteria. A review of failure criteria of fibrous composite materials was given by Echaabi et al. (1996). The maximum stress and strain criteria belong to the first category, and they are simple to apply and can tell the mode of failure, but they neglect the stress interaction. An interactive criterion such as Tsai-Wu, Hoffman, or Hill includes stress interaction in the failure mechanism, but it does not tell the mode of failure, and it requires some efforts to determine parameters such as F_{12} in the Tsai-Wu criterion. Among others, Hashin (1980) provided a three-dimensional failure criterion, which includes fiber tension, fiber compression, matrix tension, and matrix compression. This criterion not only considers the stress interaction, but also provides the failure mode. Therefore, it is widely used (Spottswood and Palazotto 2001; Kroll and Hufenbach 1997) and is adopted in this study. However, Hashin (1980) did not specify the delamination criterion, which becomes significant when the laminate fails due to interlaminar shear failure. This issue was recently addressed by Elawadly (2003). Lee (1982) proposed a delamination mode in his 3D analysis, and it is adopted in this study as an addition to Hashin's failure criterion.

Most commonly used FE models are 2D (Kim et al. 1996) and 3D analyses (Reddy and Reddy 1993). For 2D analysis, based on plane stress assumption, the transverse shear stresses, σ_{13} and σ_{23}, and normal stress, σ_{33}, are neglected. As a result, the failure mode of delamination cannot be considered. Progressive failure 3D analysis was successfully developed by several researchers. However, the disadvantage is apparent. Take a 32-layer laminate as an example; the element will be expanded 32 times more than that in 2D modeling, resulting in challenging work for both modeling and computation, which hampers its use for a parametric study. Therefore, it is the objective of this study to develop a model that uses a 2D element and can still predict the delamination failure. Since σ_{33} is negligible considering the thickness-to-length ratio for each layer, only σ_{13} and σ_{23} can be considered for delamination. In ABAQUS (2002), transverse shear stresses are not readily available in the output stress components for a shell element. Instead, they are stored in the result file as TSHR13 and TSHR23. Therefore, a user-defined subroutine is first employed to retrieve the transverse shear stresses from the result file. Combining with another subroutine to implement the failure criterion, the progressive failure analysis can be carried out.

3.2.4.2 Testing Method

A lot of tests have been carried out for failure evaluations of laminates. Standardized test methods (ASTM designations) were adopted in most studies. Cui et al. (1992) compared three- and four-point bending tests both analytically and experimentally. They concluded that in all three-point bending tests, damage was observed under the loading roller in addition to the interlaminar shear failure, while in four-point bending tests, only interlaminar shear failure was observed. Kim and Crasto (1992) carried out a series of tests on a novel miniature sandwich specimen developed to measure composite compressive strength. The mini-sandwich beam specimens consisted of thin composite skins on both sides of a core made of materials similar to matrix resin. The advantage of this method was that it can avoid the premature buckling failure, and they concluded that the compressive strength determined in their study was approximately equal to the tensile strength. But the sandwich panel fabrication was a two-step process and required more time and effort than conventional testing methods. Grief and Chapon (1993) conducted three-point bending tests on laminated composite beams and attempted to predict successive failures. Five composite laminate types were used with different lay-ups. Fiber breakage, matrix damage, and delamination were observed during the test. They tried to use total ply discount failure analysis, that is, after a ply failure the analysis was repeated for a new laminate, in which the stiffness of a failed ply was set to nearly zero, to predict subsequent failures. However, their analytical predictions did not match the experimental results. Lopez-Anido et al. (1995) performed three-point bending tests, both flatwise (out-of-plane) and edgewise (in-plane), on rectangular lay-up angle ply (±45°) beam elements. They concluded that the analysis based on the computation of the apparent lamina moduli provided a lower bound, and that based on plane strain assumptions represented an upper bound for the beam stiffness. The threshold aspect ratio that limits the range of application of various analytical methods was provided. Barbero et al. (1999) developed a fixture for testing compressive strength of coupon samples and pultruded structural shapes. Using this fixture, splitting at the end of the sample was prevented while reducing stress concentration at the ends, yielding compression failures within the center section of specimens. All the fiber reinforcements of structural shapes (Continuous Strand Mat [CSM], ±45°, and roving) were tested individually and combined to support the development of a simple model for compressive strength of structural shapes. Waas and Schultheisz (1996) provided a good review of experimental studies on compressive failure of composites. The factors affecting the compressive strength, such as matrix effects, interface effects, void content, etc., were discussed in detail through experimental results. They also correlated compressive strength with other properties and recommended testing techniques that may provide further insight into the mechanisms that control composite compressive failures, including methods such as microscopic observation, sensitive interferometry, and acoustic emission. Unlike bending and compression, tension tests are less reported due to their easy implementation.

For HFRP sandwich panels, the face laminate may be subjected to tensile, compressive, or bending forces depending on the loading conditions, where compressive force is more critical. Therefore, it is necessary to evaluate the strength properties of face laminate through a combination of compressive and bending tests, and the test results can be used to verify the accuracy of the proposed progressive failure model described in the preceding section.

3.3 Out-of-Plane Compression

3.3.1 Introduction

A combined analytical and experimental study of an FRP sandwich panel under out-of-plane compression is presented in this section (Davalos and Chen 2005). Two analytical models, corresponding to pure compression and elastic buckling failure, respectively, are provided first. The sandwich panel consists of top and bottom laminated facesheets bonded to the honeycomb core, which extends vertically between facesheets. The facesheet and core are attached by contact molding and are therefore not rigidly connected. Thus, the buckling problem can be described as the instability of an FRP core panel with two rotationally restrained loaded edges. An elastic restraint coefficient is introduced to quantify the bonding layer effect between the facesheet and core, and a simple and relatively accurate test method is proposed to obtain the restraint coefficient experimentally. By solving a transcendental equation, the critical compression buckling stresses are obtained, and a simplified expression to predict buckling strength is formulated in terms of the elastic restraint coefficient. The analytical solution is verified by FE analysis. Compression tests were carried out to evaluate the effect of the bonding layer thickness and core thickness, and the experimental results correlate closely with analytical and FE predictions. A parametric study is conducted to study the core aspect ratio effect on the buckling load. Finally, design equations are provided to calculate the compressive strength.

3.3.2 Analytical Models

Based on the literature review in Section 3.2, we conclude that there are two failure modes for HFRP sandwich panels under out-of-plane compression: pure compression and buckling. Correspondingly, two models are provided.

3.3.2.1 Pure Compression Failure

For this case, the nominal failure load can be calculated as

$$F_c = f_c \times A_c \tag{3.1}$$

where f_c is the material compressive strength of ChSM, and A_c is the total in-plane area of the core walls.

3.3.2.2 Buckling of Plate with Partially Constrained Loaded Edges

3.3.2.2.1 Analytical Model

The local buckling of core panels under uniformly distributed compression loading is analyzed in this section. Clearly, the core flat panels are more sensitive to buckling than the sinusoidal panels (Figure 3.1). Therefore, the problem can be simplified as the buckling response of the flat panel under in-plane compression. As the flat panel extends along the length of the core, it is reasonable to assume the connection edge between the flat panel and sinusoidal panel to be simply supported, as the natural location of a contra-flexure point. The boundary conditions are shown in Figure 3.2. Two edges parallel to the loading direction are simply supported and the other two loaded edges are partially constrained.

The governing differential equation for buckling of a symmetric anisotropic plate under in-plane axial loading is expressed as

$$D_{11}\frac{\partial^4 w}{\partial x^4} + 4D_{16}\frac{\partial^4 w}{\partial x^3 \partial y} + 2D_{12}\frac{\partial^4 w}{\partial x^2 \partial y^2} + 4D_{66}\frac{\partial^4 w}{\partial x^2 \partial y^2}$$

$$+ 4D_{26}\frac{\partial^4 w}{\partial x \partial y^3} + D_{22}\frac{\partial^4 w}{\partial y^4} + N_y\frac{\partial^2 w}{\partial y^2} = 0 \qquad (3.2)$$

where D_{ij} ($i, j = 1, 2, 6$) are the plate bending stiffness coefficients, N_y is the in-plane uniformly distributed compressive stress resultant, and $w(x, y)$ is the buckled shape function of the plate. If the balanced symmetric condition is

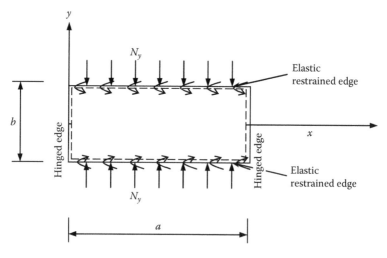

FIGURE 3.2
Boundary condition of FRP plate.

considered and no bending-twisting coupling exists, then (3.2) can be simplified as

$$D_{11}\frac{\partial^4 w}{\partial x^4} + 2D_{12}\frac{\partial^4 w}{\partial x^2 \partial y^2} + 4D_{66}\frac{\partial^4 w}{\partial x^2 \partial y^2} + D_{22}\frac{\partial^4 w}{\partial y^4} + N_y\frac{\partial^2 w}{\partial y^2} = 0 \quad (3.3)$$

Considering the boundary condition in Figure 3.2, we can assume the shape function to be

$$w = y\sin\frac{n\pi x}{a} \quad (3.4)$$

Then (3.3) can be further simplified to

$$D_{22}y^{(4)} + \left[N_y - (2D_{12} + 4D_{66})(\frac{n\pi}{a})^2\right]y'' + (\frac{n\pi}{a})^4 D_{11}y = 0 \quad (3.5)$$

Introducing the following coefficients as

$$\alpha = \frac{D_{12} + 2D_{66}}{D_{22}} \quad \beta = \frac{D_{11}}{D_{22}} \quad 2\mu^2 = (\frac{a}{n\pi})^2\frac{N_y}{D_{22}} \quad (3.6)$$

Equation (3.5) becomes

$$y^{(4)} + 2(\mu^2 - \alpha)(\frac{n\pi}{a})^2 y'' + (\frac{n\pi}{a})^4 \beta y = 0 \quad (3.7)$$

Apparently (3.7) is a typical fourth-order differential problem, and the corresponding characteristic equation is

$$r^4 + 2(\mu^2 - \alpha)(\frac{n\pi}{a})^2 r^2 + (\frac{n\pi}{a})^4 \beta = 0 \quad (3.8)$$

The final form of the solution to (3.7) depends on the value of $(\mu^2 - \alpha)^2 - \beta$. We can assume

$$(\mu^2 - \alpha)^2 - \beta = 0 \quad (3.9)$$

Substituting (3.6) into (3.9), we can get

$$\left[(\frac{a}{n\pi})^2\frac{N_y}{2D_{22}} - \frac{D_{12} + 2D_{66}}{D_{22}}\right]^2 - \frac{D_{11}}{D_{22}} = 0 \quad (3.10)$$

Solving for (3.10), we have

$$N_y = \frac{2n^2\pi^2}{a^2}\left[\sqrt{D_{11}D_{22}} + (D_{12} + 2D_{66})\right] \tag{3.11}$$

This is a well-known expression for the critical local buckling strength of a simply supported plate with n half waves in the x direction (Reddy 1999). For the problem considered in this study, for a given n, we always have $N_{cr} \geq N_y$, and therefore we always have

$$(\mu^2 - \alpha)^2 - \beta \geq 0 \tag{3.12}$$

Then the four roots of (3.8) are the complex numbers

$$r = \pm\frac{k_1 n\pi}{a}i; \quad \pm\frac{k_2 n\pi}{a}i \tag{3.13}$$

where k_1, k_2, and k_3 are defined as

$$k_1 = \sqrt{\mu^2 - \alpha + k_3}$$

$$k_2 = \sqrt{\mu^2 - \alpha - k_3} \tag{3.14}$$

$$k_3 = \sqrt{(\mu^2 - \alpha)^2 - \beta}$$

Then the solution for (3.7) takes the form

$$w(x,y) = \sin\frac{n\pi x}{a}(C_1\cos\frac{k_1 n\pi y}{a} + C_2\sin\frac{k_1 n\pi y}{a} + C_3\cos\frac{k_2 n\pi y}{a} + C_4\sin\frac{k_2 n\pi y}{a}) \tag{3.15}$$

As indicated in Figure 3.2, the origin of the coordinates x and y is located at the mid-point of the left edge. Assuming equal elastic constraint on both loaded edges, the deflection function of w is a symmetric function of y when the plate reaches the critical buckling load. Therefore, (3.15) reduces to

$$w(x,y) = \sin\frac{n\pi x}{a}(C_1\cos\frac{k_1 n\pi y}{a} + C_3\cos\frac{k_2 n\pi y}{a}) \tag{3.16}$$

The boundary conditions can be described as

$$w_{y=\pm h/2} = 0 \tag{3.17}$$

The rotational angle is assumed to be proportional to the edge moment,

$$M_{y=\pm h/2} = -\bar{\zeta}\bar{\varphi} \tag{3.18}$$

where $\bar{\varphi}$ is the rotation of the plate along the edges $y = h/2$.

Based on the constitutive equation of a laminated panel, and considering $(\partial^2 w / \partial x^2)_{y=\pm h/2} = 0$, the moment M_y is expressed as

$$M_{y=\pm h/2} = -D_{22}\left(\frac{\partial^2 w}{\partial y^2}\right)_{y=\pm h/2} \tag{3.19}$$

Combining (3.18) and (3.19), we have

$$\bar{\varphi} = \frac{D_{22}}{\bar{\zeta}}\frac{\partial^2 w}{\partial y^2} \tag{3.20}$$

A nondimensional factor or coefficient of elastic restraint (CER) is defined as

$$\zeta = -\frac{D_{22}}{\bar{\zeta}}\frac{2}{h} \tag{3.21}$$

Considering $\bar{\varphi} = \partial w / \partial y$, the boundary condition along the edges $y = \pm h/2$ becomes

$$\frac{\partial w}{\partial y} = \frac{h}{2}\zeta\frac{\partial^2 w}{\partial y^2} \tag{3.22}$$

The buckled shape function of (3.16) in combination with (3.17) and (3.22) results in homogeneous equations in terms of two constants C_1 and C_3. When the determinant of the coefficient matrix equals zero, the buckling criterion for a plate under equal elastic constraint on both loaded edges is established as

$$\begin{vmatrix} \cos\dfrac{k_1 n\pi h}{2a} & \cos\dfrac{k_2 n\pi h}{2a} \\[2ex] -\dfrac{k_1 n\pi}{a}\sin\dfrac{k_1 n\pi h}{2a} - \zeta\dfrac{h}{2}(\dfrac{k_1 n\pi}{a})^2\cos\dfrac{k_1 n\pi h}{2a} & -\dfrac{k_2 n\pi}{a}\sin\dfrac{k_2 n\pi h}{2a} - \zeta\dfrac{h}{2}(\dfrac{k_2 n\pi}{a})^2\cos\dfrac{k_2 n\pi h}{2a} \end{vmatrix} = 0 \tag{3.23}$$

Furthermore, (3.23) is simplified to a transcendental equation as

$$k_1\sin\frac{k_1 n\pi h}{2a}\cos\frac{k_2 n\pi h}{2a} - k_2\sin\frac{k_2 n\pi h}{2a}\cos\frac{k_1 n\pi h}{2a} + \zeta\frac{n\pi h}{a}k_3\cos\frac{k_1 n\pi h}{2a}\cos\frac{k_2 n\pi h}{2a} = 0 \tag{3.24}$$

A Fortran program is compiled to solve this equation. As pointed out by Reddy (1999), for a simply supported plate under uniaxial compression, the buckling load is a minimum when the half wave along the unloaded direction is 1. The theory also applies to this model. It is found out that $n = 1$ always gives the minimum buckling load, while the number of half waves along the other direction can be calculated by the program for a corresponding buckling load.

3.3.2.2.2 Verification with FE Simulation

To verify the model derived in Section 3.3.2.2.1, both (3.24) and the FE method are used to predict the local buckling strength of the core panel under out-of-plane compression. The structure is a typical single cell of the honeycomb sandwich structure. This cell is 10.16×10.16 cm square and 5.08 cm deep, and the core thickness is $t = 2.29$ mm, as shown in Figure 3.3. Table 3.1 lists stiffness properties of the core wall.

ABAQUS (2002) is adopted for FE analysis, and FEMAP (2001) is used for the pre- and postprocessing. The modeling of the complex shape of a sinusoidal wave is accomplished by exporting the geometry from AUTO-CAD. The core walls are modeled with a four-node shell element, S4. The global element size is chosen as 5.08 mm. It was checked through a convergence study that the mesh used provided accurate values. In the FE analysis, the CER introduced in Section 3.3.2.2.1 is adopted to account for the bonding

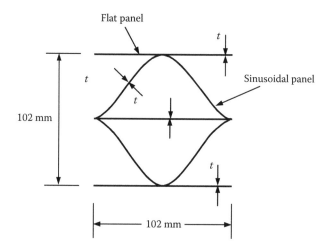

FIGURE 3.3
In-plane core specimen dimensions.

TABLE 3.1

Properties of the Core Material

Ply name	Orientation	E_1 (GPa)	E_2 (GPa)	G_{12} (GPa)	G_{23} (GPa)	v_{12}	v_{23}
Core	Random	11.79	11.79	4.21	2.97	0.402	0.388

layer effect, and a spring model is used to simulate the elastic constraint. Each node between the facesheet and the core is duplicated, and six spring elements, representing the constraints in six directions, are placed in between. The normal spring stiffness is set to be a very large value. This dummy value prevents the core from detaching from the facesheet. The rotational stiffness is varied to represent the relative constraining condition, corresponding to a particular elastic restraint coefficient. An eigenvalue analysis is carried out, where the load corresponding to the first buckling mode is considered as the buckling load. It is shown that the buckling load is dependent on CER, and by varying CER, denoted as ζ, we can plot the buckling load as shown in Figure 3.4, with the first buckling mode from the FE analysis illustrated in Figure 3.5.

Solving for (3.24), we can obtain the buckling load N_y for the flat panel in the cell, which is 10.16 cm wide and 5.08 cm deep. If the compressive stress is assumed to be evenly distributed for the whole structure, multiplying N_y by the total length of all the core walls, we can plot the buckling load vs. elastic restraint coefficient in Figure 3.4, from which it is shown that the analytical model fits the FE result quite well. When the coefficient of elastic restraint is assumed to be very large, which approaches a hinged connection, (3.24) gives the result of $\mu^2 = 3.125$, and substituting this value into (3.6), N_y can be calculated as

$$N_y = 2\mu^2 \frac{\pi^2 D_{22}}{a^2} = \frac{6.25\pi^2 D_{22}}{a^2} \tag{3.25}$$

which corresponds to the solution given by Reddy (1999) for a plate under in-plane compression with four sides simply supported, leading to the solution

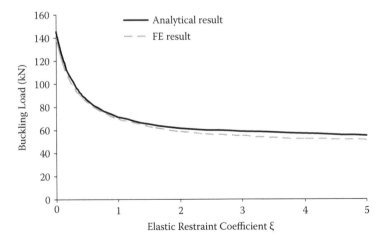

FIGURE 3.4
Buckling load vs. elastic restraint coefficient.

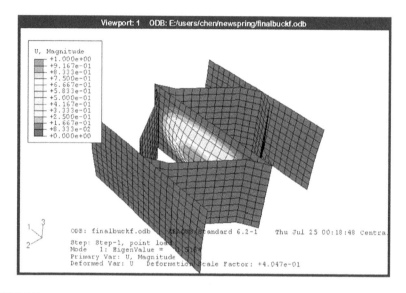

FIGURE 3.5
The first buckling mode for clamped condition.

$$N_{cr} = \frac{\pi^2 D_{22}}{a^2}(\frac{a}{h}+\frac{h}{a})^2 = \frac{6.25\pi^2 D_{22}}{a^2} \qquad (3.26)$$

which is identical to (3.25), thus indirectly verifying the accuracy of the above formulation.

For a given CER ζ, we can get the buckling load correspondingly from the curves shown in Figure 3.4. To simplify this procedure, we provide an explicit expression to predict the buckling load, which can also act as a design equation. Previous investigations (Qiao et al. 2001) showed that the buckling load vs. ζ curve shown in Figure 3.4 can be fitted using the following equation:

$$\frac{F_{cr} - F_{cr}^\infty}{F_{cr}^0 - F_{cr}^\infty} = \frac{1}{p\zeta^q + 1} \qquad (3.27)$$

where F_{cr}^0 and F_{cr}^∞ are critical loads corresponding to the hinged ($\zeta = \infty$) and clamped ($\zeta = 0$) boundary conditions, respectively. They can be obtained from the analytical solution and FE analysis and are listed in Table 3.2. The parameters p and q can be determined from (3.24) by a regression technique, and the results from both the FE and analytical solutions are given in Table 3.2.

To further verify the analytical solution, panels composed of $2 \times 2 = 4$ (20.32×20.32 cm) and $3 \times 3 = 9$ (30.48×1 30.48 cm) cells with the same core height are analyzed under compressive load. For simplicity, only the two extreme cases of hinged and clamped conditions are illustrated, and the

TABLE 3.2

Comparison between FE and Analytical Result

	F_{cr}^{∞} (kN)	F_{cr}^{0} (kN)	p	q
FE result	45.36	140.34	2.94	1.01
Analytical result	50.04	145.62	3.48	1.07

TABLE 3.3

Comparison between FE and Analytical Result for Multicell Panel

		F_{cr}^{∞} (kN)	F_{cr}^{0} (kN)
2 × 2 = 4 cells	FE result	186.91	541.62
(203 × 203 mm²)	Analytical result	183.82	533.35
3 × 3 = 9 cells	FE result	397.81	1,093.90
(305 × 305 mm²)	Analytical result	397.97	1,161.64

results given in Table 3.3 show that the analytical solution correlates well with FE results.

3.3.3 Experimental Investigation

To further study the behavior of sandwich panels under out-of-plane compression, an experimental investigation was carried out by two types of tests, stabilized and bare compression, to correspondingly achieve pure compression and buckling failures, as described in Section 3.3.2.

3.3.3.1 Naming Conventions

Throughout this study, the naming conventions are defined in Figure 3.6, where B and C represent, respectively, number of chopped strand mat (ChSM) bonding layer and core thickness, and different values for i and j correspond to different nominal weights of the ChSM.

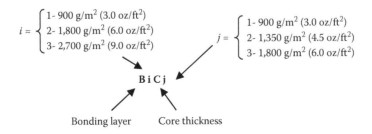

$$i = \begin{cases} 1 - 900 \text{ g/m}^2 \ (3.0 \text{ oz/ft}^2) \\ 2 - 1,800 \text{ g/m}^2 \ (6.0 \text{ oz/ft}^2) \\ 3 - 2,700 \text{ g/m}^2 \ (9.0 \text{ oz/ft}^2) \end{cases} \qquad j = \begin{cases} 1 - 900 \text{ g/m}^2 \ (3.0 \text{ oz/ft}^2) \\ 2 - 1,350 \text{ g/m}^2 \ (4.5 \text{ oz/ft}^2) \\ 3 - 1,800 \text{ g/m}^2 \ (6.0 \text{ oz/ft}^2) \end{cases}$$

B i C j

Bonding layer   Core thickness

FIGURE 3.6
Naming conventions.

3.3.3.2 Test Description

The specimen is a typical single cell cut from the sandwich structure, which represents the weakest part of the structure when under compression. It is 10.16×10.16 cm and 5.08 cm deep, as shown in Figure 3.3. To assess the effect of bonding layers and minimize the influence of the other layers of the facesheet, only three layers are selected for the facesheet, as shown in Figure 3.7. The thickness of bonding layers is varied from one bonding layer to three bonding layers, and the core thickness of specimens is varied from one to two core thickness. The constituent materials of the facesheet are given in Figure 3.7. The properties of the constituent materials are provided in Table 3.4, and Table 3.5 lists the properties of each component material.

Two cases of compression tests were carried out. For the first case, an elastic pad was placed between the loading block and the specimen; this method is known as bare compression test. For the second case, the specimen was bonded to top and bottom steel plates, and the load was applied directly over the steel plate; this method is called stabilized compression test. The bare compression test is more representative of actual patch loading conditions. The stabilized compression test is intended to minimize the buckling effect and induce primarily compression failure.

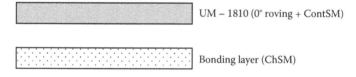

UM – 1810 (0° roving + ContSM)

Bonding layer (ChSM)

FIGURE 3.7
Lay-up of facesheet.

TABLE 3.4

Properties of Constituent Materials

Material	E (GPa)	G (GPa)	ν	ρ (g/cm³)
E-glass fiber	72.4	28.8	0.255	2.55
Polyester resin	5.06	1.63	0.3	1.14

TABLE 3.5

Layer Properties of Face Laminate and Core Materials

Ply Name	Ply Type	Nominal Weight (g/m²)	Thickness (mm)	V_f
Each bonding layer	ChSM	915.5	2.08	0.1726
UM1810	0°	610.3	0.635	0.3774
	CSM	305.2	0.335	0.3582
Core	ChSM	1373.4	2.28	0.2359

FIGURE 3.8
Compression test setup.

All tests were carried out according to ASTM standards (see Figure 3.8). They were performed in a universal testing machine with an 889.6 kN capacity. A load cell was placed between the loading block and the specimen to record the load, and linear variable differential transducers (LVDTs) were used to record the displacements. Four strain gages were bonded at the midheight of the core to obtain compressive strains, two on the sinusoidal wave panel and two on the side flat panel (Figure 3.3). The load was controlled at such a rate that the failure occurred within 3 to 6 min.

3.3.3.3 Test Results and Discussion

3.3.3.3.1 Bare Compression Test

When the load is applied to the specimen, both side flat panels bend outwards, and this deformation can be interpreted as a geometric imperfection. As the load increases, the specimens with distinct bonding layers display different behaviors. For B1C2, the side panels buckle and delaminate from the specimen well before ultimate failure occurs. While for other types, the side panels do not delaminate. For all specimen types, upon sudden crushing of the side panel, the specimen does not fail immediately but continues to carry load for several event failures, until collapse of the specimen. A typical failure mode is shown in Figure 3.9. The maximum loads for specimens with distinct bonding layers are shown in Figure 3.10, and the average value and standard deviation for six samples each are given in Table 3.6, which shows that the magnitudes of failure loads are in the same order as the number of bonding layers and core thickness; i.e., the specimen with three bonding layers is much stronger than that with one bonding layer, and

FIGURE 3.9
Bare compression test specimen.

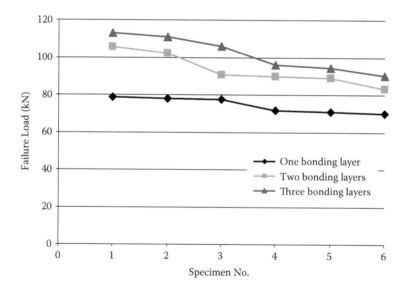

FIGURE 3.10
Failure load for bare compression test.

TABLE 3.6

Average Value and Standard Deviation of Failure Load for Bare Compression Tests

	B1C2	B2C2	B3C2	B3C1
Average value (kN)	74.60	93.46	101.86	31.74
Standard deviation (kN)	3.89	8.47	9.43	3.45

the specimen with two core thicknesses is stronger than that with one core thickness, clearly showing that the bonding layer effect and core thickness play an important role on the failure load. Figure 3.11 shows the load-displacement curve. Figure 3.12 shows the transverse strain vs. load curve for

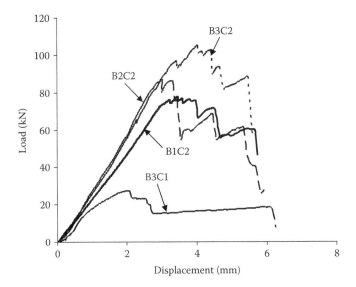

FIGURE 3.11
Load-displacement curve for bare compression test.

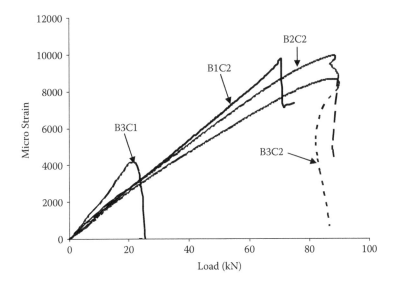

FIGURE 3.12
Strain-load curve for bare compression test.

the sinusoidal panel. As the elastic pad is placed between the loading block and the specimen, this displacement does not represent the actual deformation of the specimen. However, from these figures we can conclude that the specimen exhibits an approximate linear behavior up to failure.

3.3.3.3.2 Stabilized Compression Test

In this test, all three types of samples show the same failure mode. They all fail by crushing of the core panels. The sinusoidal wave panel fails first, followed by the crushing of the remaining components of the core, where the failure mode is shown in Figure 3.13. No apparent damage can be observed prior to ultimate failure.

The failure loads for three specimens each are given in Table 3.7, which shows much higher values than what we obtained for the bare compression tests. Figure 3.14 shows a typical load-displacement curve for the inside sinusoidal wave panels, and Figure 3.15 gives the strain-load curve. Again, we can see that the specimens follow a nearly linear behavior until failure occurs.

FIGURE 3.13
Stabilized compression test specimen.

TABLE 3.7

Average Value of Failure Load for Stabilized Compression Test

	B1C2	B2C2	B3C2
Average value (kN)	155.54	163.07	177.22
Range (kN)	148.73–162.36	156.71–171.83	158.13–201.59

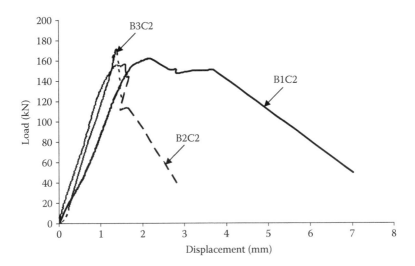

FIGURE 3.14
Load-displacement curve for stabilized compression test.

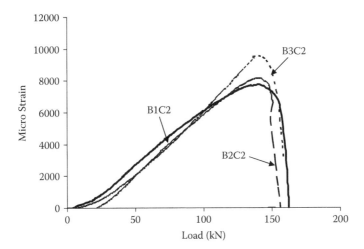

FIGURE 3.15
Strain-load curve for stabilized compression test.

3.3.3.4 Discussion of Experimental Results

From previous studies, we can estimate the compressive strength for the ChSM to be about 153.1 MPa for the present fiber volume fraction (Barbero et al. 1999). From the same test method, as will be described in Section 3.5, the compressive strength is found to be 95.64 kN, as shown in Appendix 3.A. Halpin and Kardos (1978) suggested a model to predict the compressive strength for ChSM using a pseudoisotropic lamination method. Following their method, if the compression failure strains for the equivalent unidirectional composite are assumed to be $\varepsilon_{1c} = 0.015$ and $\varepsilon_{2c} = 0.006$, we can obtain the stress-strain curve to failure as given in Figure 3.16. Therefore, it is reasonable to assume the compressive strength for ChSM material to be 148.2 MPa. The total in-plane area of the core walls with two core thicknesses is 13.74 cm². Then, the nominal failure load can be calculated as

$$F_c = f_c \times A_c = 148.2 \times 13.74 = 203.49 \, \text{kN} \qquad (3.28)$$

where f_c is the compressive strength of ChSM and A_c is the total in-plane area of the core walls.

The stabilized test gives the failure load ranging from 148.7 to 201.2 kN. If the unevenly distributed load effect is considered, we can conclude that the stabilized compression test results in a typical compression failure. For the bare compression test, the failure load is much lower than the nominal compressive load. This indicates that local buckling probably occurs before the structure gains its maximum compressive strength. Once the local buckling occurs, the buckled parts of the specimen lose their function and the compressive load is redistributed among the other parts. Finally, the structure fails in compression or a combination of bending and compression.

The two types of tests resulted in two distinct failure modes. Buckling occurred for the bare compression test, while the stabilized compression test induced material compression failure. As a matter of interest, the two failure modes were the same as those reported by Zhang and Ashby (1992)

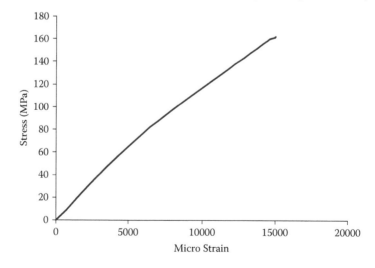

FIGURE 3.16
Stress-strain curve for ChSM.

under out-of-plane compression. As expected, the failure loads of stabilized compression tests are much higher than those for the bare compression tests.

3.3.4 FE Analysis

The same technique as described in Section 3.3.2 is used to carry out the FE analysis to correlate with test results. As discussed in the previous section, the stabilized compression test leads to compression failure, and the bare compression test is initiated by local buckling. Therefore, two types of analyses are carried out: static analysis and buckling analysis.

3.3.4.1 Load-Strain Curve

A linear static analysis is used for the stabilized compression test; a buckling analysis is carried out for the bare compression test. As the bending of the side panels is observed in the bare compression test, geometric imperfection is included in the model for the bare compression test to account for this deformation. The core wall thickness is used as a scale factor for geometric imperfection: 0.5 t for the side panels. After extracting the fourth eigenmode (Figure 3.17), the modified Riks method is used in the analysis (ABAQUS 2002). As the compressive load in the test is applied through a rigid loading block, the facesheet should displace downward at the same rate. Thus, a multiple-point constraint (MPC) condition is used to allow the nodes in the same horizontal plane to move at the same displacement. Figures 3.18 and 3.19 show

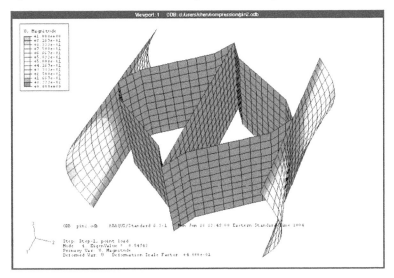

FIGURE 3.17
Imperfection mode.

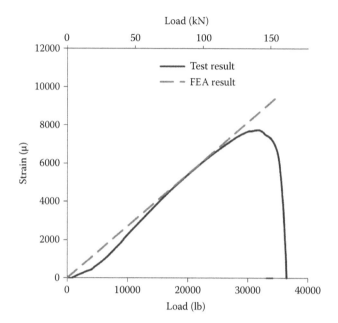

FIGURE 3.18
Load-strain curves for stabilized compression test.

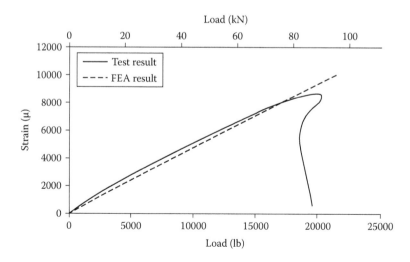

FIGURE 3.19
Load-strain curves for bare compression test.

comparisons of FE analysis results and test results for strain-load responses, showing good correlation between the two results.

3.3.4.2 Analysis Results and Discussion

The FE results indicate that the buckling load is 135.08 kN for the clamped condition and 48.13 kN for the hinged condition. The failure load of bare compression tests falls within this range, which indicates that the actual connection lies between simply supported and fully restrained conditions.

CER is dependent on the constrain element between the facesheet and core, such as the bonding layer thickness and core stiffness. If this coefficient can be determined, the local buckling strength can be computed. Therefore, a necessary step is to independently define the elastic restraint coefficient, which will be given in Section 3.3.5 through a cantilever plate test.

3.3.5 Determination of the Coefficient of Elastic Restraint

As pointed out earlier, the facesheet and core are not rigidly connected, and therefore CER is defined to quantify the degree of connectivity at the interface. To determine this coefficient, a testing method is developed in this section. The test setup is schematically shown in Figure 3.20, with the core wall embedded into the facesheet. Figure 3.21 displays a photograph of the test setup.

If the connection is rigid, considering the line load acting at the end of this cantilever plate and neglecting the shear deformation of the thin plate, the deflection at the end for a rigid boundary condition is given as

$$\Delta_1 = \frac{wb^3}{3D_{22}} \tag{3.29}$$

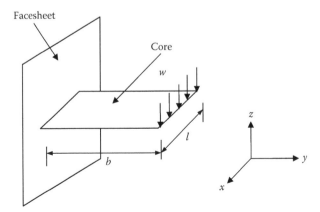

FIGURE 3.20
Test setup to determine the elastic restraint coefficient.

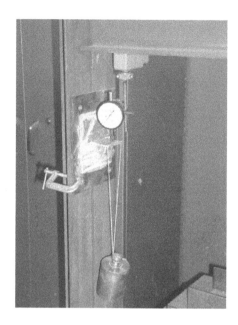

FIGURE 3.21
Photo of test setup.

where w is the distributed line load acting at the end of the plate. However, as the core element is not rigidly connected, there is a rotation at the connection, which can be calculated as

$$\overline{\varphi} = \frac{M}{\overline{\zeta}} = \frac{wb}{\overline{\zeta}} \tag{3.30}$$

The relative deflection at the end of the plate corresponding only to this rotation is

$$\Delta_2 = \overline{\varphi}b \tag{3.31}$$

where b is the length of the panel, as shown in Figure 3.20.
    Then the total deflection becomes

$$\Delta = \Delta_1 + \Delta_2 = \frac{wb^3}{3D_{22}} + \overline{\varphi}b \tag{3.32}$$

Following the same procedure, if w is acting at the mid-span of the plate, the deflections at the end of the plate, for rigid connection and due to the relative rotation, can be respectively calculated as

$$\Delta_1' = \frac{5wb^3}{48D_{22}} \tag{3.33}$$

$$\Delta_2' = \frac{wb}{2\bar{\zeta}} = \frac{\bar{\varphi}b}{2} \tag{3.34}$$

$$\Delta' = \Delta_1' + \Delta_2' = \frac{5wb^3}{48D_{22}} + \frac{\bar{\varphi}b}{2} \tag{3.35}$$

Solving simultaneously for (3.32) and (3.35), we obtain

$$\bar{\varphi} = \frac{3.2\Delta' - \Delta}{0.6b} \tag{3.36}$$

Substituting (3.36) into (3.30), using (3.32), and based on the definition of ζ in (3.21), we can obtain the coefficient of restraint ζ through some simple transformations as

$$\zeta = \frac{16\Delta' - 5\Delta}{12\Delta - 24\Delta'} \tag{3.37}$$

This expression (3.37) shows that this coefficient is only related to the two deflections, irrespective of the dimensions of the plate and the applied load. Thus, the accuracy of this testing method depends only on the measurement of tip displacements for the two load cases. We can test the validity of (3.37) by considering two extreme cases. If the connection is completely rigid, only the deflection corresponding to a rigid end is present as $\Delta = 3.2\Delta'$, resulting in $\zeta = 0$. While for a hinged connection, the flexural deflection is negligible compared with the tip displacement, due to the hinge rotation, which becomes $\Delta = 2\Delta'$, and results in $\zeta = \infty$. These results correspond to the range of values defined previously for ζ.

Tests were carried out for the three cases of distinct bonding layer thickness. The specimens were cut from the same samples as used in the compression tests described above, with l and b (Figure 3.20) both equal to 5.08 cm. A standard weight of 2 kg was used to apply the load at both the end and mid-span of the plate, and a dial gage with a precision of 0.00254 mm was used to measure the displacement at the end of the plate. The test results are listed in Table 3.8.

TABLE 3.8

Average Value of CER with $b = 5.08$ cm

	One Bonding Layer	Two Bonding Layers	Three Bonding Layers
Average value	0.84	0.41	0.29
Range	0.81–0.87	0.40–0.41	0.27–0.33

3.3.6 Comparisons of Test Results with Analytical and FE Predictions

Using the CER value obtained from the cantilever plate test described above in (3.24), we can predict the buckling load. Two sets of p and q are used, one from the analytical solution and the other from the FE analysis, and their corresponding critical loads for hinged and fixed conditions, as listed in Table 3.2. The results are summarized in Table 3.9, showing good correlations of test results with analytical and FE predictions.

3.3.7 Parametric Study

In practice, it is common to vary the core height to meet the requirement for the panel depth. Using the analytical model derived and the CER obtained, we can carry out a parametric study by varying the core height for the specimen studied. The critical buckling stress vs. core height curve is illustrated in Figure 3.22 for $a = 10.16$ cm, from which it can be seen that the buckling stress is quite sensitive to variation of cell height up to about 10.16 cm, and

TABLE 3.9

Comparison of Analytical, FE, and Test Results for Buckling Load

	B1C2	B2C2	B3C2
Analytical result (N)	74,641	90,971	99,809
FE result (N)	72,764	88,600	96,864
Test result (N)	74,596	93,457	101,864

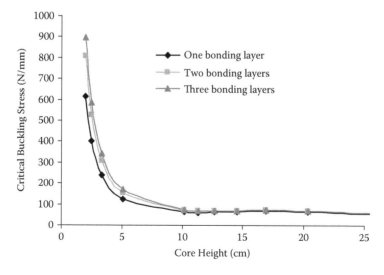

FIGURE 3.22
Critical buckling stress vs. core height.

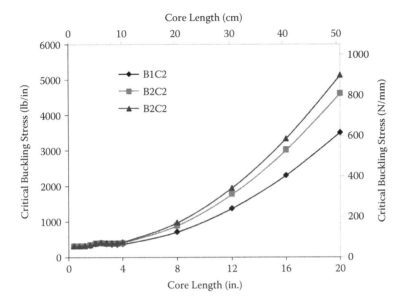

FIGURE 3.23
Critical buckling stress vs. length of flat panel.

within this range there is a notable difference among the buckling stresses for a different number of bonding layers. The buckling stress decreases as the core height increases, and the stress reaches nearly a plateau when the core height is higher than 20.32 cm. Beyond this limit point the bonding layer thickness does not affect the result much. The reason for this behavior is that when the aspect ratio of height over width is less than 1, the number of the half waves parallel to the loading direction is one, and therefore the boundary condition will affect the buckling load. But as core height increases, more half waves along the loaded direction will result, and in this case, the buckling load will be determined by the wavelength in between the two loaded edges. As a result, the boundary condition does not affect the buckling load much. If we keep the height fixed, we can find the relationship between the buckling stress and length of the flat panel, as shown in Figure 3.23. The buckling stress increases as the length increases, and it is anticipated that when the length increases to infinity, i.e., the aspect ratio approaches zero, the plate will not buckle. Clearly, Figures 3.22 and 3.23 also illustrate the relationship between the critical buckling stress and the aspect ratio of the core wall.

Multiplying the buckling stress by the total core wall length, the buckling load vs. core height curve is given in Figure 3.24 for a = 10.16 cm case for different bonding layers. For a given height, we can easily find the buckling load for a single cell from these curves.

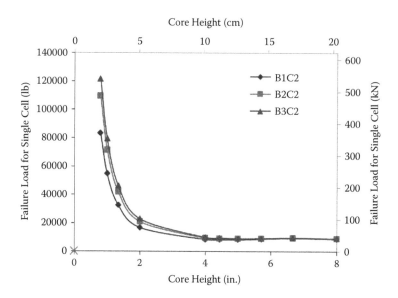

FIGURE 3.24
Buckling load vs. core height.

3.3.8 Design Equations

Design equations can be developed based on the analytical model derived above. Only the two-core thickness case, i.e., $t = 2.286$ cm, which is most commonly used, is considered, while the other core thicknesses can be included following the same manner.

Considering buckling failure, the three curves shown in Figure 3.24 can be fitted using the following equation:

$$F = 4.4482 \times (A_1 e^{[-h/(B_1 \times 25.4)]} + A_2 e^{[-h/(B_2 \times 25.4)]} + F_0) \qquad (3.38)$$

The parameters corresponding to each bonding layer are listed in Table 3.10. It is noted that (3.38) gives the failure load for a single cell, and if it is divided by the in-plane area for a single cell, which is $10.16 \times 10.16 = 103.23$ cm² for this case, the buckling compressive strength can be obtained.

TABLE 3.10

Parameters for Design Equation

	A_1	B_1	A_2	B_2	F_0
One bonding layer	957,515	0.2363	124,742	0.7464	8,081
Two bonding layers	87,639	1.0105	95,4711	0.2917	8,136
Three bonding layers	1,038,189	0.2985	88,384	1.0765	8,152

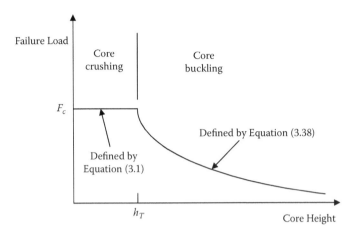

FIGURE 3.25
Design diagram.

TABLE 3.11

Transition Height

	One Bonding Layer	Two Bonding Layers	Three Bonding Layers
h_T	32 mm	36 mm	38 mm

Either (3.1) or the stabilized compression test can be used to find the failure load corresponding to core crushing, where the average value of $F_c = 164.58$ kN from the test results is adopted herein. Based on the failure modes of core crushing and buckling, we can propose a design equation as shown in Figure 3.25, where h_T is the height where the failure mode transits from core crushing to core buckling, as listed in Table 3.11.

3.3.9 Concluding Remarks

Two analytical models, corresponding to pure compression and elastic buckling failure, respectively, are provided for panels subject to out-of-plane compression. A combined analytical and experimental study of elastic buckling analysis is given for FRP panels with elastic restraint at the loaded edges. By solving a transcendental equation, the critical compression buckling stresses are obtained. An elastic restraint coefficient is employed to quantify the elastic restraint effect, namely, the bonding layer effect. Buckling loads are calculated in terms of the elastic restraint coefficient. The analytical predictions are verified by FE analysis. The compression test is carried out to study the behavior of sandwich panels under out-of-plane compression. A cantilever plate test is conducted to define the coefficient of elastic restraint. Both the closed-form solution and FE analysis are used to predict the buckling load for the given test samples, and the results are in good correlation. A parametric study is carried out to study the aspect ratio effect on the buckling load. Based on the study presented in this section, the following conclusions can be drawn:

1. The closed-form solution derived in this chapter can predict the buckling strength of a plate with partially restrained loaded edges. Unlike existing solutions for eigenvalue analysis, where the number of half waves should be predefined when calculating the buckling load, this solution can give the minimum buckling load and the corresponding number of half waves. The accuracy is verified by FE analysis and experimental results.

2. Typically there are two failure modes for HFPR core under out-of-plane compression, buckling, and pure compression failure. The buckling load is sensitive to the bonding layer effect. Specimens with three bonding layers fail at a higher load than those with one bonding layer. While for pure compression failure, the failure loads are not affected much by the number of bonding layers.

3. Bonding layer effect can be interpreted through a coefficient of elastic restraint (CER). It plays an important role on the buckling behavior. However, rigid connection is commonly used in the analysis of sandwich structures, corresponding to $\zeta = 0$ in this study. It is shown that this assumption can lead to a significant error if the aspect ratio is within a certain limit.

4. CER can be predicted using the testing method provided. Only two deflections are required to calculate this coefficient. Thus, the accuracy is increased. This method, together with the interface shear test and interface tension test, can be used as the criteria to define the bonding quality of a given connection.

5. A parametric study is carried out by varying the core height. The result indicates that if the core height is relatively low, there is a notable difference of the buckling stress for different numbers of bonding layers. The buckling stress decreases as the core height increases and reaches nearly a constant value once the core height reaches a certain limit. The buckling load is no longer sensitive to the bonding layer effect at this stage.

6. Practical design equations are provided to calculate the compressive strength, and these guidelines can be easily implemented.

The bonding layer effect not only affects the buckling load, but also influences the behavior of the sandwich panels under out-of-plane shear. This will be presented in the following section. The method described in this chapter can be further applied to other structures with elastic restraint at the loaded edges, like the web buckling in FRP or even steel girders.

3.4 Out-of-Plane Shear

3.4.1 Introduction

A combined analytical and experimental study of FRP sandwich panel under out-of-plane shear is presented in this section (Chen and Davalos 2007). Analytical models are provided for delamination considering skin effect, shear crushing, and shear buckling. Two factors are addressed that contribute to the skin effect: shear warping and bending warping. A closed-form solution, based on proper description of a displacement field at the interface, is derived considering shear warping. The accuracy of this method is verified by FE results. The FE model is then applied to study bending warping effect. The stiffness and the stress distribution subject to skin effect are presented. Critical structural sections are identified, and suggestions for future design considerations are given. Based on the stress distribution, design formulas for delamination and shear failure are presented. The Rayleigh–Ritz method is employed to study the shear buckling of core panels with two sides elastically restrained. Four-point bending tests were carried out according to ASTM standards to study shear strength and shear stiffness of the core materials. The number of bonding layers and core thickness were varied to study their effects on strength. Two types of beam samples were manufactured by orienting the sinusoidal wave: (1) along the length (longitudinal) and (2) along the width (transverse). Different failure modes were observed for different types of specimens. Design equations are developed to predict the failure load due to different failure modes, and good correlations are obtained with test results.

3.4.2 Analytical Model Including Skin Effect

3.4.2.1 Origin of Skin Effect

As shown in Chapter 2, the lower bound of the transverse shear stiffness for this sinusoidal core can be provided neglecting skin effect. There is no study on accurate description of the transverse shear stiffness and stress distribution, partly due to the complex displacement field, especially for curved panels, such as the sinusoidal panel in this study. Chen and Davalos (2005) pointed out that the displacement field in cell walls for sandwich core can be described by two distinct modes:

1. Directly at the face-core interface, if facesheet is assumed to be rigid, which is reasonable considering the stiffness ratio between the facesheet and core. This is defined by displacement compatibility, where strain transformation can be used to find the relationship between local and global strain.
2. At a position sufficiently far away from the interface, i.e., such as at the mid-depth where the effect of rigid facesheet dissipates. This is defined by force equilibrium.

Therefore, the purpose of the analysis is to find a displacement field that can accurately describe these two distinct modes and the transition field in between. In order to achieve this, the displacement field at the interface has to be properly described first. A basic assumption for all previous studies on the equivalent properties of the sandwich core is that the cell walls predominantly carry load through membrane strain, and that the bending forces in the cell walls are neglected. However, the bending effect should play a role when defining the shear stiffness and shear distribution, especially when the core height is low. Therefore, we believe that shear and bending warping effects are better descriptors of these phenomena, where shear warping corresponds to the assumption adopted in the previous studies, and the bending warping describes the additional bending effect offered by the skin. Furthermore, it is found out that shear warping corresponds to cases with hinge connection between facesheet and core, and when both warping effects are considered, it corresponds to a rigid connection. The actual cases usually lie in between these two conditions. Detailed descriptions of skin effect are as follows.

3.4.2.2 Skin Effect

A unit cell of a honeycomb sandwich panel and its dimensions are shown in Figures 3.26 and 3.27, respectively. Two factors may contribute to skin effect: shear and bending warping.

3.4.2.2.1 Shear Warping

The resulting distributed shear flow for a typical cell and its representative volume element (RVE) are shown in Figures 3.28 and 3.29. The equilibrium equation and compatibility condition for a longitudinal wave configuration without considering skin effect can be written as

$$4t \int_0^l \tau_2 \cos\theta \, ds + 2ta\tau_1 = Ha\tau_{xz} \tag{3.39}$$

$$2 \int_0^l \tau_2 \, ds / G_{12} = \tau_1 a / G_{12} \tag{3.40}$$

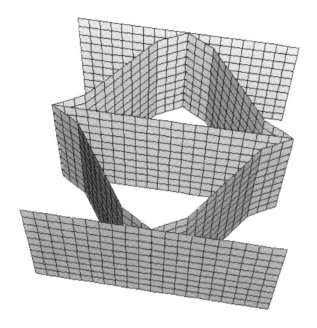

FIGURE 3.26
Unit cell.

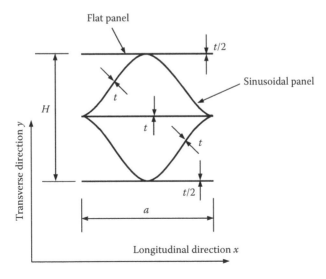

FIGURE 3.27
Dimensions of a unit cell.

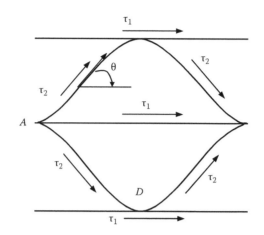

FIGURE 3.28
Shear flow in the unit cell.

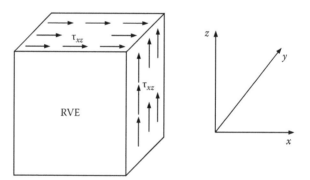

FIGURE 3.29
Shear flow in the RVE.

where

$$2\int_0^l \cos\theta \, ds = a,$$

G_{12} is the material shear modulus, l is the curved panel length, and t, a, and H are defined in Figure 3.27. Solving for (3.39) and (3.40), we have

$$\tau_1 = \frac{H \displaystyle\int_0^l ds}{at + 2t \displaystyle\int_0^l ds} \tau_{xz} \tag{3.41}$$

$$\tau_2 = \frac{H}{2t + 4t \int_0^l ds/a} \tau_{xz} \tag{3.42}$$

From (3.41) and (4.4), we have

$$\tau_1 / \tau_2 = 2 \int_0^l ds/a \tag{3.43}$$

and correspondingly, we can obtain

$$\gamma_1 / \gamma_2 = 2 \int_0^l ds/a \tag{3.44}$$

where γ_1 and γ_2 are the shear strains in the flat and curved panels, respectively. Apparently, the flat panel will deform along a straight line, while the curved panel deforms along a curved shape as shown in Figure 3.30 (only half of the top curve is shown). However, in most practical cases, the face and the core are constrained so that they remain essentially plane during deformation. Therefore, to compensate the deformation shown in Figure 3.30, the shear warping occurs at the top and bottom of a curved panel. The expression of shear warping will be given in Section 3.4.2.3.

3.4.2.2.2 Bending Warping

Pure shear strain in the curved wall will induce a displacement in the x direction, as shown in Figure 3.31. However, if we assume core-facesheet is rigidly connected, the rotation at the top and bottom of the core is constrained, resulting in a deformed shape, as shown in Figure 3.31. This phenomenon

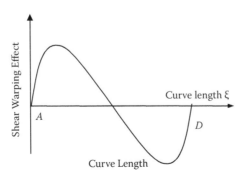

FIGURE 3.30
Shear warping (plan view).

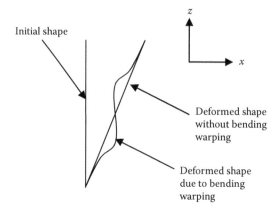

FIGURE 3.31
Bending warping (elevation view).

can be termed bending warping. Apparently, an additional moment at both the top and bottom will result due to this effect.

It should be noted that although both effects are local, they can significantly affect the stress distribution at both the top and bottom of the core, i.e., the interfacial stresses, as discussed below.

3.4.2.3 Theoretical Analysis

Consider the element ABCD in Figure 3.32, which is cut from the unit cell shown in Figure 3.27, subject to a shear strain γ. The equilibrium equations for the stresses acting on the $\xi\eta$ plane in the absence of body forces are

$$\partial\sigma_\xi / \partial\xi + \partial\sigma_{\xi\eta} / \partial\eta = 0 \tag{3.45}$$

$$\partial\sigma_{\xi\eta} / \partial\xi + \partial\sigma_\eta / \partial\eta = 0 \tag{3.46}$$

Considering the stress-strain relationship, Chen and Davalos (2005) further reduced (3.45) and (3.46) into the form (see Appendix 3.B)

$$G(\partial^2 v / \partial\xi^2) + E'(\partial^2 v / \partial\eta^2) = 0 \tag{3.47}$$

where $E' = E/(1 - v^2)$. The stress components can be defined as

$$\sigma_\eta = E'(\partial v / \partial\eta) \tag{3.48}$$

$$\tau_{\xi\eta} = G(\partial v / \partial\xi) \tag{3.49}$$

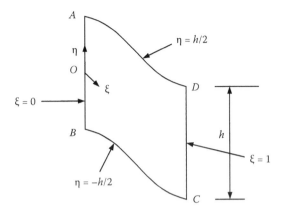

FIGURE 3.32
Model used for theoretical analysis.

$$\sigma_\xi = vE'(\partial v / \partial \eta) \tag{3.50}$$

Equations (3.47) to (3.50) act as the basis for this analytical study. The boundary conditions considering shear warping are

$$v(0, \eta) = v(l, \eta) = 0$$
$$v(\xi, \eta) = v(\xi, -\eta) \tag{3.51}$$
$$v(\xi, h/2) = \varphi(\xi)$$

where l is the curved panel length and h is the height. Then, $\varphi(\xi)$, caused by shear warping as shown in Figure 3.30, can be defined as

$$\varphi(\xi) = \gamma_1 x(\xi) - \gamma_2 s(\xi) \tag{3.52}$$

where $x(\xi)$ = length of flat panel, and $s(\xi)$ = length of curved panel.
    The solution of (3.47) can be described using Fourier series as

$$v(\xi, \eta) = \sum_{n=1}^{\infty} \left[\frac{2/l}{\cosh(\frac{n\pi h}{2l\mu})} \varphi_n \cosh(\frac{n\pi\eta}{l\mu}) \sin(\frac{n\pi\xi}{l}) \right] \tag{3.53}$$

where

$$\mu = \sqrt{E'/G} \tag{3.54}$$

$$\varphi_n = \int_0^l \varphi(\xi) \sin(\frac{n\pi\xi}{l}) d\xi \tag{3.55}$$

The normal stress σ_η can be obtained using (3.48) as

$$\sigma_\eta(\xi,\eta) = E'(\partial v/\partial \eta) = E' \sum_{n=1}^{\infty} \left[\frac{2n\pi}{l^2 \mu \cosh(\frac{n\pi h}{2l\mu})} \varphi_n \sinh(\frac{n\pi\eta}{l\mu}) \sin(\frac{n\pi\xi}{l}) \right] \tag{3.56}$$

Equation (3.49) gives

$$\tau(\xi,\eta) = G(\partial v/\partial \xi) = G \sum_{n=1}^{\infty} \left[\frac{2n\pi}{l^2 \cosh(\frac{n\pi h}{2l\mu})} \varphi_n \sinh(\frac{n\pi\eta}{l\mu}) \cos(\frac{n\pi\xi}{l}) \right] \tag{3.57}$$

The normal stress σ_ξ can be obtained using (3.50).
   Next, the total strain energy is defined as

$$U = \int \frac{\tau^2}{2G_{12}} dV + \int \frac{\sigma^2}{2E} dV \tag{3.58}$$

in order to obtain the equivalent shear modulus G_{xz}:

$$G_{xz} = \frac{2U}{V\gamma^2} \tag{3.59}$$

where U is total strain energy, V is the volume corresponding to the RVE, and γ is the shear strain applied to the structure, which is equal to γ_1 in value. The above equations can be incorporated into any mathematical software, such as MATHCAD.

3.4.2.4 Description of FE Model

The FE method is employed to verify the analytical model derived in Section 3.4.2.3. A unit cell of a honeycomb sandwich panel and its dimensions are shown in Figure 3.26. Due to the symmetric structure, we can further reduce the cell into a quarter cell, as shown in Figure 3.33. This quarter cell will be used in the FE analysis. Based on symmetry, the thickness is $t/2$ for the flat panels and t for the sinusoidal panel. The height of the core is half of the unit cell dimensions. The dimensions and properties of the core materials are listed in Table 3.12.

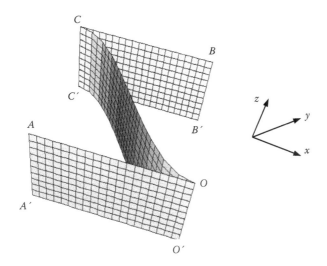

FIGURE 3.33
Model used for FE analysis.

TABLE 3.12

Properties of Core Mat

H, mm	a, mm	t, mm	E (GPa)	G (GPa)	ν
102	102	2.29	11.79	4.21	0.402

In the FE analysis, all the nodes at the top face translate at a uniform displacement in the x direction. The shear force can be computed by summing up the reaction force along the x direction for all the nodes at the top. Thus, the equivalent shear stiffness G_{xz} and shear stress distribution can be obtained. The boundary conditions are listed in Table 3.13, where CB/AO is defined according to the restraint condition assumed. In particular, free, pinned, and fixed boundary conditions correspond to free, hinge, and rigid connections.

3.4.2.5 Application

3.4.2.5.1 Equivalent Shear Stiffness

Figure 3.34 plots shear stiffness vs. aspect ratio, where aspect ratio is defined as h/a. The lower bound is given in Chapter 2 without considering skin effect. From Figure 3.34, we can observe that the analytical solution, considering shear warping, corresponds to hinge connection. There is a significant difference between hinge and rigid cases when the aspect ratio is low, whereas all the solutions approach the lower bound value as aspect ratio increases. This proves that, as pointed out by several researchers (Xu et al. 2001), the skin effect is localized, and its effect on stiffness, which is a global parameter,

TABLE 3.13

Boundary Conditions of FE Model

	u_x	u_y	u_z
OO′/BB′	Free	0	0
CC′/AA′	Free	0	0
CB/AO	Constant	—	—
C′B′/A′O′	0	0	Free

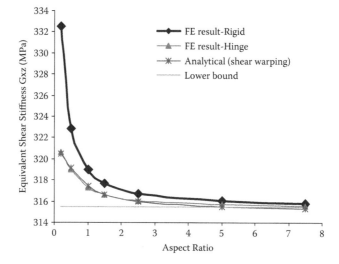

FIGURE 3.34
Stiffness vs. aspect ratio.

is negligible when the core is high enough. However, the skin effect does affect the stresses significantly, as will be discussed next.

Using the regression technique, we can express the transverse shear stiffness as

$$G_{xz} = 315.85 + 247717.08e^{-\frac{R-0.1127}{0.2136}} + 5.73e^{-\frac{R-0.1127}{1.2393}} \tag{3.60}$$

for rigid connection, where R is the aspect ratio, and

$$G_{xz} = 315.15 + 15.15e^{-\frac{R-0.1144}{0.6438}} + 0.99e^{-\frac{R-0.1144}{9}} \tag{3.61}$$

for pinned connection from FE result, and

$$G_{xz} = 313.88 + 5.23e^{-\frac{R-0.1113}{0.7987}} + 0.99e^{-\frac{R-0.1113}{20}} \tag{3.62}$$

for pinned connection from analytical result.

Therefore, the stiffness is a function of the aspect ratio R. The expression (3.62) acts as a lower limit, and can be used in the analysis and design for safety considerations.

It is also interesting to point out that, as concluded by Kelsey et al. (1958), the theory of minimum potential energy, a kinematically compatible uniform strain field, gives an upper bound, and the theory of complementary energy, a statically compatible uniform stress field, gives a lower bound, corresponding to infinitely large and zero skin effects, respectively. Voigt and Reuss (see Christensen 1991) expressed this theory in parallel and series models as

$$\frac{1}{2}\frac{\sigma_{ij}^2}{C_{ij}}V \leq \sum_{k=1}^{n}(U_b + U_s + U_a)_k \tag{3.63}$$

$$\frac{1}{2}C_{ij}\varepsilon_{ij}^2V \leq \sum_{k=1}^{n}(U_b + U_s + U_a)_k \tag{3.64}$$

where k accounts for individual substructures in the RVE, and U_b, U_s, and U_a are, respectively, the strain energies related to bending, shear, and axial responses. Expressions (3.63) and (3.64) define, respectively, the conditions of lower and upper bounds for stiffness. As shown in Chapter 2, these two equations give an upper (328.05 MPa) and lower (315.6 MPa) bound. Comparing these two values with the results shown in Figure 3.34, we can note that the lower bound still applies, while the upper bound does not exist anymore. This, once again, can be explained by the fact that bending warping was neglected in previous studies, and therefore the stiffness was underestimated.

3.4.2.5.2 Stress Distribution

Both the analytical method and the FE method are applied to a particular example; the core panel height is $h = 5.08$ cm. The results are listed in Figures 3.35 to 3.38.

3.4.2.5.2.1 Shear Warping   In order to illustrate the shear warping effect, all stresses are plotted in Figure 3.35 for both flat and curved panels. From Figure 3.35, we can note that, in the flat panel, the stress distribution is not affected, and the shear stress remains constant and the normal stress is essentially zero. While for the curved panel, the shear warping effect is significant, the minimum shear stress occurs at the center of the curved panel, and the distribution of normal stress is as shown in Figure 3.35.

Figure 3.36 plots the stress distributions along the top of the curved panel, as calculated from both analytical and FE results, showing good correlations. The same phenomenon can be observed for stress distribution along the height at the panel intersection, as shown in Figure 3.37. This proves the

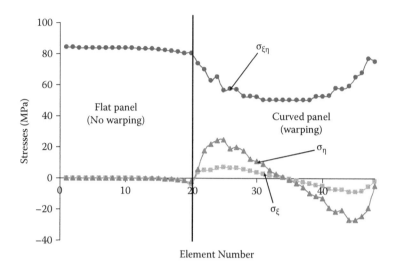

FIGURE 3.35
Stress distribution.

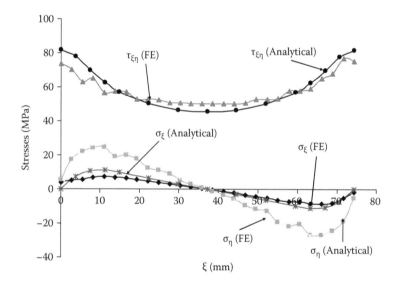

FIGURE 3.36
Comparisons between FE and analytical result for stress distribution (along length).

accuracy of the analytical method for predicting the behavior of the curved panel under shear warping.

3.4.2.5.2.2 Bending Warping   Figure 3.38 shows the stress distributions for the curved panel assuming a rigid connection between core and facesheet, from which we can note that, due to the bending effect explained above, the normal

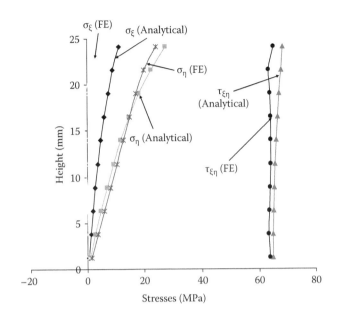

FIGURE 3.37
Comparisons between FE and analytical result for stress distribution (along height).

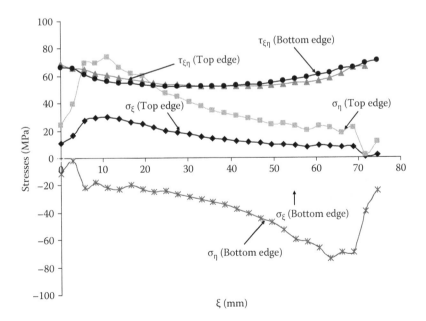

FIGURE 3.38
Stress distribution with bending warping based on the FE analysis.

stress is no longer constant along the thickness t of the core wall. Stresses, positive at the top and negative at the bottom of the core wall section, result from the extra bending moment due to the rotation incompatibility. The shear stress distribution is also affected, the value of which decreases compared to the hinge connection, illustrating the benefits that rigid connection can offer.

3.4.2.5.2.3 Discussion   From Figure 3.36, we observe that the ratio between the interfacial shear stress (81.84 MPa) and the interfacial tensile stress (27.11 MPa) is approximately 3. If bending warping is considered, the tensile stress can be even larger than the shear stress, as shown in Figure 3.38. Based on the results from the flatwise tension test and interfacial shear test, Wang (2004) pointed out that a typical interfacial shear strength (12.06 MPa) is four to five times the interfacial tensile strength (2.76 MPa). Therefore, it is reasonable to assume that the delamination is caused by the tensile force at the interface (corresponding to mode I fracture). The tensile force can be used to predict the onset of the delamination. Once the crack occurs, there is a stress singularity at the crack tip, and the fracture mechanics method should be used to predict crack growth, using parameters such as fracture toughness, crack length, J-integral, etc.

3.4.2.6 Parametric Study

Using the closed-form solution derived in this chapter, a parametric study is carried out for the interfacial normal stress, S_{22}, at the panel intersection under a shear strain of 0.02, as shown in Figure 3.39, from which we can

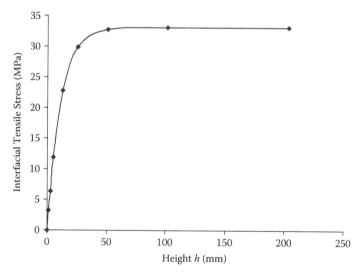

FIGURE 3.39
Normal interfacial stresses vs. height.

observe that S_{22} increases as the aspect ratio increases, and reaches a constant value beyond a certain limit, for instance, $h \approx 2$ in. for this case. The curve shown in Figure 3.39 can be fitted using

$$S_{22} = 33.12 - 33.128e^{-[0.09123h)]^{1.0221}}$$

(3.65)

Equation (3.65) is based on a shear strain of 0.02. For a unit shear strain, (3.65) can be normalized as

$$S_{22} = \gamma(1655.78 - 1655.78e^{-[0.09123h)]^{1.0221}})$$

(3.66)

where γ is the shear strain.

3.4.2.7 Summary

In this section, the skin effect, composed of shear and bending warping, on the behavior of HFRP sandwich sinusoidal core panels is for the first time investigated. An analytical solution is given for shear warping, and FE analyses are carried out for both shear and bending warping cases. It is concluded that:

1. The analytical solution can successfully predict the behavior of curved panels considering shear warping, which is verified by FE results.
2. Skin effect includes two parts: shear and bending warping. Shear warping corresponds to cases with hinge connection between facesheet and core, and when both warping effects are considered, it corresponds to a rigid connection. Actual cases lie between these two conditions.
3. The skin effect is a localized phenomenon. The lower bound of the equivalent stiffness can thereby be adopted if the aspect ratio is high enough. However, it can significantly affect interface stress distribution, yielding a coupled stress state for the curved panel, where the normal stress may even be larger than the shear stress. This indicates, unlike the common belief that only shear stress occurs when the structure is under pure shear force, that tensile force at the interface arises, making it a potentially critical component. Therefore, special considerations are suggested for design purposes.
4. The skin effects described herein only affect the stress distribution of the curved panel and have no effect on the flat panel. This effect on the stress distribution becomes less significant in the area away from the interface.

5. Practical formulas to calculate equivalent shear stiffness and interfacial normal stress are provided. Together with flatwise tension test results, they can be used for failure predictions, as will be shown in Section 3.4.7.

3.4.3 CER Effect on Shear Stiffness and Interfacial Shear Stress Distribution

In Section 3.3, we concluded that the CER effect will greatly affect the buckling strength of core panels under out-of-plane compression. It is interesting to find out that this effect can also contribute to the shear stiffness and interfacial shear distribution. The same model shown in Section 3.4.2.4 is employed, and spring elements, as described in Section 3.3, are placed at the interface to simulate the partially constrained condition.

Figure 3.40 plots the FE results for G_{xz} vs. CER curve, from which we can note that completely rigid boundary conditions (CER = 0) correspond to the largest value of G_{xz}. The FE results fall within the range of the lower (315.44 MPa) and upper (328.05 MPa) bound solutions given by Davalos et al. (2001). However, the absolute maximum difference is 1.2%, which is negligible. From FE analysis, the shear stress contour indicates that the shear stress at top nodes is uniform except in the area adjacent to the connection of the flat and sinusoidal wave panels, where the shear stress decreases. Therefore, this nearly uniform stress can represent the interfacial shear

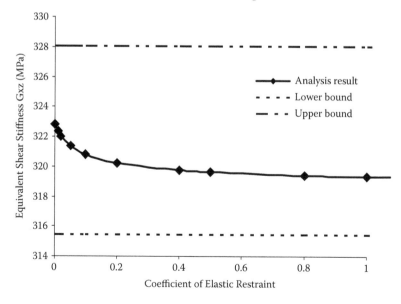

FIGURE 3.40
CER effect on transverse shear stiffness.

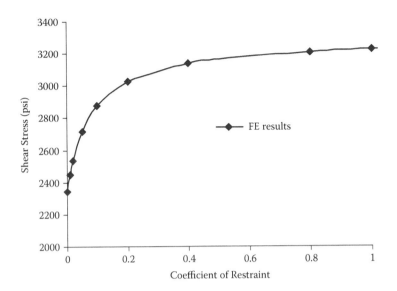

FIGURE 3.41
CER effect on interfacial shear stress.

stress. Figure 3.41 displays the relationship between CER and interface shear stress, from which we can see that shear stress increases as CER increases, with maximum values for near hinged conditions (CER ≥ 1.0). Therefore, the shear stress corresponding to the hinged condition can be adopted to predict shear crushing failure for design purposes.

3.4.4 Shear Buckling

The core may buckle due to shear loading if the core is deep and thin. The solution for shear buckling is provided. Following the approach given by Qiao et al. (2001) and considering the boundary conditions shown in Figure 3.42, the following first variation of the total potential energy equation is used to define the problem:

$$\int_0^a \int_0^h (D_{11}\frac{\partial^2 w}{\partial x^2}\delta\frac{\partial^2 w}{\partial x^2} + D_{12}\frac{\partial^2 w}{\partial x^2}\delta\frac{\partial^2 w}{\partial y^2} + D_{12}\delta\frac{\partial^2 w}{\partial x^2}\frac{\partial^2 w}{\partial y^2}$$

$$+ D_{12}\frac{\partial^2 w}{\partial x^2}\delta\frac{\partial^2 w}{\partial y^2} + D_{22}\frac{\partial^2 w}{\partial y^2}\delta\frac{\partial^2 w}{\partial y^2} + 4D_{66}\frac{\partial^2 w}{\partial x\partial y}\delta\frac{\partial^2 w}{\partial x\partial y}$$

$$+ N_{xy}\frac{\partial w}{\partial x}\delta\frac{\partial w}{\partial y} + N_{xy}\delta\frac{\partial w}{\partial x}\frac{\partial w}{\partial y})dxdy \qquad (3.67)$$

$$+ \int_0^a [\bar{\zeta}(\frac{\partial w}{\partial y})_{y=0}\delta(\frac{\partial w}{\partial y})_{y=0} + \bar{\zeta}(\frac{\partial w}{\partial y})_{y=h}\delta(\frac{\partial w}{\partial y})_{y=h}]dx = 0$$

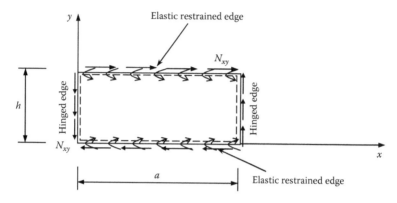

FIGURE 3.42
Boundary condition of FRP plate.

Using the Rayleigh–Ritz method, the displacement $w(x, y)$ that satisfies the boundary conditions (excluding the case when the boundary conditions are clamped, (i.e., $\zeta = \infty$), can be defined as

$$w = \sum_{i=1}^{m} \sum_{j=1}^{n} A_{ij} \sin \frac{i\pi x}{a} \sin \frac{j\pi y}{h} \tag{3.68}$$

Substituting (3.68) into (3.67), a typical eigenvalue problem results. The results of the eigenvalues are in the form of pairs of ± quantities, which means there is no direction requirement for the shear stress. The smallest eigenvalue can be taken as the critical shear stress resultant. Figure 3.43 shows the critical shear stress of shear buckling for one bonding layer and two core thicknesses. An asymptotic value can be assumed for the aspect ratio $h/a > 5$, when a sufficient number of terms (e.g., $m = n = 6$) is included (Qiao et al. 2001). The critical buckling stress for different bonding layers is shown in Figure 3.44, from which we can observe that the difference for the bonding layers effect on shear buckling capacity is negligible.

The curves shown in Figure 3.44 can be fitted using

$$N = 0.175(A_1 e^{-\frac{R}{t_1}} + A_2 e^{-\frac{R}{t_2}} + N_0) \tag{3.69}$$

where N is the critical shear stress and R is the aspect ratio. The parameters corresponding to different bonding layers are listed in Table 3.14. The shear stress can be expressed as

$$\tau = \frac{N}{t} = 0.175 \times (\frac{A_1}{t} e^{-\frac{R}{B_1}} + \frac{A_2}{t} e^{-\frac{R}{B_2}} + \frac{N_0}{t}) \tag{3.70}$$

where t is the core wall thickness.

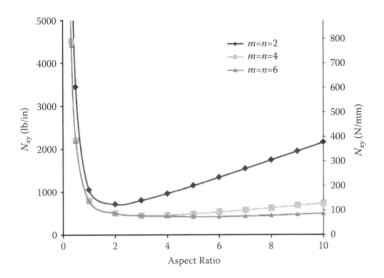

FIGURE 3.43
Critical shear stress vs. aspect ratio for one bonding layer.

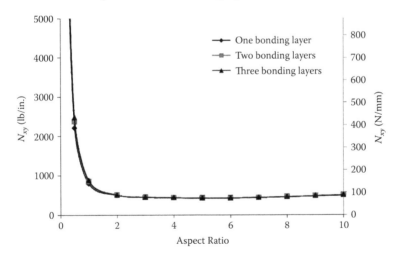

FIGURE 3.44
Critical shear stress vs. aspect ratio for different bonding layers.

TABLE 3.14

Parameters for Design Equation

	A_1	B_1	A_2	B_2	N_0
One bonding layer	2,103	0.5326	34,611	0.1388	448
Two bonding layers	2,661	0.5097	37,093	0.1355	449
Three bonding layers	3,015	0.4970	38,734	0.1339	450

3.4.5 Proposed Method to Predict Failure Load

It is shown (Caprino and Langella 2000) that if the core Young's modulus is negligible with respect to the facing elastic modulus, and the facing thickness is small compared to the height of the core, the transverse shear stress field in the core is practically uniform. Therefore, the following basic assumptions are adopted in this model:

1. Transverse shear stress is carried by the core only.
2. Transverse shear stress is uniformly distributed along the core height.
3. The structure is considered to fail once the transverse shear stress exceeds the critical shear strength, either shear strength of the material or buckling strength.

3.4.5.1 Core-Face Delamination

From the discussion above, we can observe that when under pure shear force, tensile force at the interface arises, making it a potentially critical component. Therefore, special considerations are suggested for design purposes. Based on the analytical model derived in this section, we can propose the following design guidelines using the failure criterion of maximum stress:

1. For a given loading condition, calculate shear strain based on the equivalent shear modulus by (3.62).
2. Find the interfacial tensile stress from (3.66) using shear strain calculated from step 1.
3. Compare this interfacial tensile stress with the interfacial tensile strength from the flatwise tension test.

This method will provide a conservative result since (1) the shear stiffness corresponding to the hinged connection between the core and facesheet is adopted, and (2) it is shown (Wang 2004) that a crack is initiated when the interface traction attains the interfacial strength, and the crack is advanced when the work of traction equals the material's resistance to crack propagation. Therefore, a nominal interfacial tensile strength will be used in order to propose a more reasonable criterion. The validity of the proposed method will be discussed through the correlation with four-point bending test results, as will be shown in Section 3.4.7.

3.4.5.2 Core Shear Failure and Shear Buckling

From the analysis shown above, it is found that the shear stress in the flat panel is higher than that in the curved panel. Therefore, the flat panel is more

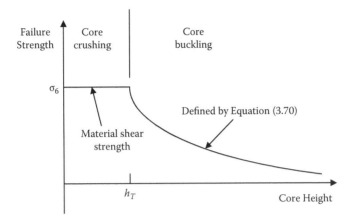

FIGURE 3.45
Design diagram.

TABLE 3.15

Transition Height

	One Bonding Layer	Two Bonding Layers	Three Bonding Layers
h_T	88 mm	94 mm	99 mm

critical when considering pure shear failure and shear buckling. Following the same method for compressive strength, we can propose a design equation as shown in Figure 3.45, where h_T is the height where the failure mode transits from core crushing to core buckling, as listed in Table 3.15.

The material shear strength can be obtained from a V-notched test (Iosipescu test, ASTM D5379-98), as shown in Appendix 3.A. The average value of five specimens is $\sigma_6 = 10{,}239$ MPa.

Following the same approach described in Section 3.3.8, the following design guidelines are proposed for the shear capacity of a flat panel:

1. For a given loading condition, calculate shear strain based on the equivalent shear modulus by (3.62).
2. Calculate shear stress in the flat panel.
3. Compare the shear stress with the strength obtained from Figure 3.45.

This method will be illustrated in Section 3.4.7.

3.4.6 Experimental Investigation

It is generally recommended that ASTM C273-00 be used for shear properties of sandwich core materials, as shown in Figure 3.46, which was also initially

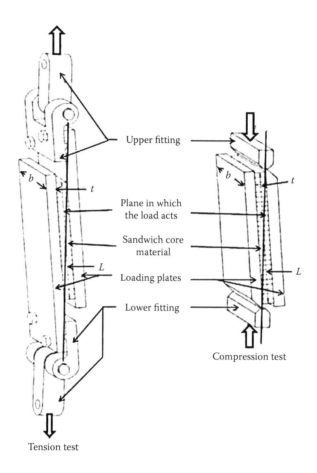

FIGURE 3.46
Plate shear specimens. (Reprinted from ASTM C273-00 (2011), copyright ASTM International, 100 Barr Harbor Drive, West Conshohocken, PA 19428. With permission. A copy of the complete standard may be obtained from ASTM International, http://www.astm.org.)

adopted in this study. It was finally abandoned because it was found from trial tests that, due to high shear strength of the core material, the delamination in the facesheet, i.e., intralaminate delamination, occurred well before the shear failure of the core material can be achieved. An alternate method, four-point bending test (ASTM C393-00), is also recommended by ASTM for the study of core shear strength and shear modulus, which was used by a lot of researchers, as shown in the literature review, and is also adopted herein.

3.4.6.1 Test Description

The dimensions of the specimen were 71.12 cm long by 10.16 cm wide by 5.08 cm deep. There were seven single cells along either the longitudinal or transverse direction, as shown in Figure 3.47. To minimize the influence of

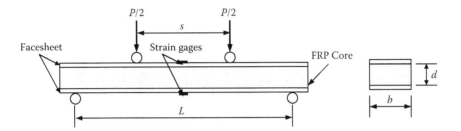

FIGURE 3.47
Test setup.

the layers of the facesheet other than the bonding layer on the strength of the specimen, only a combined 0°/continuous strand mat (CSM) layer was placed over the ChSM bonding layer, as shown in Figure 3.7. The constituent materials of the facesheet are given in Figure 3.7, and their properties are provided in Table 3.15, with the properties of each component material given in Table 3.16.

The core of the sandwich panels was embedded into the facesheet using a ChSM contact layer and resin. The number of these bonding layers was varied from one to three to study their effect on strength. Two types of beam samples were manufactured by orienting the sinusoidal wave: (1) along the length (longitudinal) and (2) along the width (transverse). All tests were carried out in accordance with ASTM standards. Figure 3.47 displays the test setup, where $L = 60.96$ cm and $s = 30.48$ cm. An external load cell was placed between the loading block and the specimen to record the load, and LVDTs were used to record the displacements. Two strain gages on the top and two on the bottom facesheets were bonded at the mid-span of the beam (Figure 3.47). The test was performed at a displacement rate of 1.524 cm/min. A photograph of the test setup is shown in Figure 3.48.

3.4.6.2 Test Results and Discussion

3.4.6.2.1 Longitudinal Test

The beams under static loadings showed nearly linear elastic behavior up to failure. The number of bonding layers affects the mechanical behavior of the specimens. For the range of one to three bonding layers, the failure of the specimens was due to a sudden debonding between the facesheet and the core material, as shown in Figure 3.49. The energy stored in the specimen was released in a relatively short time, resulting in a loud failure. For the excessive bonding layers, the facesheet did not delaminate from the core, and a typical shear failure of the core under the loading point occurred instead, as shown in Figure 3.50.

The average values of the maximum loads of three specimens for excessive bonding layers and five specimens for each of the other types are given in

Table 3.16, which shows that the magnitudes of failure loads are in the same relation as the number of bonding layers and core thickness; i.e., the specimen with three bonding layers is much stronger than that with one bonding layer, and the specimen with three core thicknesses is much stronger than that with one core thickness, clearly showing that the effect of the number of bonding layers and core thickness plays an important role on the failure load. This is due to the fact that, by increasing the number of the bonding layers and the core thickness, larger fillets of excess adhesive are formed at the honeycomb interface, and this increases the bonding area. Figure 3.51(a)

FIGURE 3.48
Photo of test setup.

FIGURE 3.49
Failure due to delamination.

FIGURE 3.50
Shear crushing of the core.

TABLE 3.16

Average Value of Failure Load for Longitudinal Samples

	B1C2	B2C2	B3C2	B2C1	B2C3	B3C1	B3C3	Excessive Bonding Layers
Average value (N)	16,681	24,176	30,159	17,637	23,509	23,620	41,413	70,460
Standard deviation (N)	890	2,180	3,737	1,668	3,559	1,824	3,292	17,726

and (b) shows the displacement at mid-span vs. load curves for specimens with two bonding layers and two core thicknesses. Figure 3.52(a) and (b) shows the load-strain curves for the same specimens. From these figures we can conclude that the specimens exhibited an approximate linear behavior up to failure.

3.4.6.2.2 Transverse Test

All types of specimens tested displayed the same failure mode. The failure in the core was initiated by debonding at the contact area between the sinusoidal panel and flat panel, as shown in Figure 3.53. The specimens continued to carry some load until the delamination between the facesheet and core material occurred. Unlike longitudinal specimens, the failure was not as sudden, and several rises and drops of load were observed during the test.

The failure loads for five specimens each are given in Table 3.17, which shows much lower values compared with what we obtained for the longitudinal samples. Therefore, the transverse specimens should not be used when

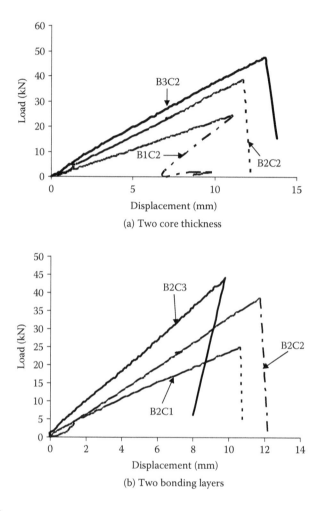

FIGURE 3.51
Load-displacement curve for longitudinal test.

high shear stresses are expected. Figure 3.54(a) and (b) shows typical load-displacement curves for specimens with two bonding layers and two core thicknesses. Figure 3.55(a) and (b) shows typical load-strain curves for the same specimens. We can observe that the specimens followed a nearly linear behavior until first failure occurred.

3.4.6.3 Summary

An investigation on the strength properties of HFRP specimens in bending is conducted through four-point bending tests. In particular, the influence of facesheet-core interface bonding effect is examined by varying the

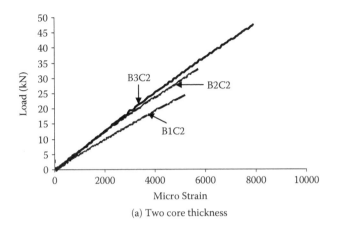

(a) Two core thickness

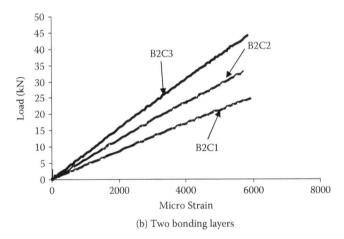

(b) Two bonding layers

FIGURE 3.52
Load-strain curve for longitudinal test.

bonding layers of the specimen. Two cases of bending tests are carried out: longitudinal and transverse bending tests. It is found that:

1. All specimens followed an approximate linear behavior prior to failure in bending. The failure load for the longitudinal specimens is much higher than that for the transverse specimens. For longitudinal samples, the specimens with excessive bonding layers failed in shear, and the other specimen types failed by debonding. All of the transverse specimens failed by debonding. Transverse-type beams should be avoided when high shear stresses are expected.

2. The failure load is sensitive to the bonding layer effect and core thickness effect. Specimens with more bonding layers and core thickness failed at a higher load than those with less bonding layer and core

FIGURE 3.53
Core separation.

TABLE 3.17

Average Value of Failure Load for Transverse Samples

	B1C2	B2C2	B3C2	B2C1	B2C3	B3C1	B3C3
Average value (kN)	5.25	6.98	11.19	5.29	7.16	6.78	11.03
Standard deviation (kN)	0.71	1.27	2.67	0.33	0.76	0.87	1.82

thickness. The failure load may vary for the same type of specimen due to the variability of bonding quality, which indicates the importance of quality control during manufacturing of the panels.

From the test result, we can observe, as expected, that the longitudinal samples are much stronger in shear than the transverse samples. The number of bonding layers and core thickness correspond clearly to the maximum strengths achieved. However, there is variability in results even for specimens with the same number of bonding layers, especially for the type with excessive bonding layer. One of the factors that may contribute to this variability is the bonding quality. For some specimens, the fillets are not well formed at the core-facesheet interface, resulting in minor cracks. This indicates the importance of quality control during the manufacturing process.

3.4.7 Correlations between Test Results and Prediction from Design Equations

For longitudinal specimens, two types of failure modes were observed, pure shear failure and delamination. In this section, the models derived in

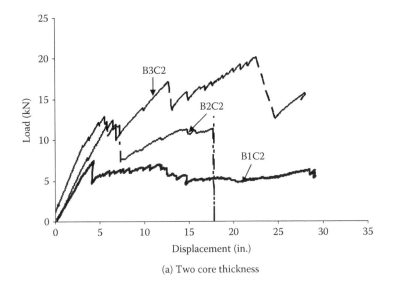

(a) Two core thickness

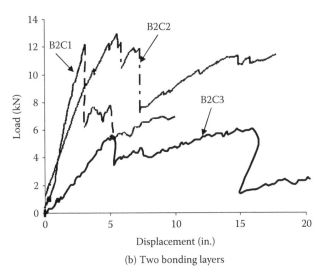

(b) Two bonding layers

FIGURE 3.54
Load-displacement curve for transverse test.

Section 3.4.5 are used to predict the failure strength corresponding to these two distinct failure modes.

3.4.7.1 Shear Failure of Flat Panel

The model described in Section 3.4.5 is applied to the longitudinal specimen with excessive bonding layers. Based on basic assumptions, we have

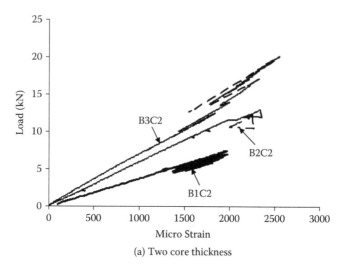

(a) Two core thickness

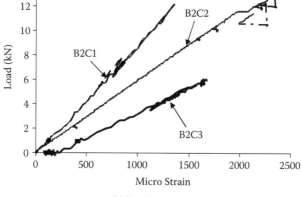

(b) Two bonding layers

FIGURE 3.55
Load-strain curve for transverse test.

$$\gamma = \frac{P}{2G_{xz}b(d+t_f)} \tag{3.71}$$

where b and d are given in Table 3.18, and G_{xz} is the equivalent shear modulus given by (3.62). For $d = 5.08$ cm, we have

$$G_{xz} = 318.06 \text{ MPa} \tag{3.72}$$

TABLE 3.18

Parameters for Sandwich Beam Specimen with Excessive Bonding Layers

E_f (×10² MPa)	E_c (×10² MPa)	b (mm)	d (mm)	t_f (mm)
133.8	5.3	114	50.8	7.22

Since the shear strain of the flat panel is the same as the global shear strain, the shear stress in the flat panel can be calculated as

$$\tau = \gamma G_{12} \tag{3.73}$$

where G_{12} is the material shear modulus from Table 3.1. The transition height h_T equals 9.75 cm for three bonding layers; therefore, $h < h_T$, and shear crushing controls. Substituting into (3.73) and (3.71), we can obtain the failure load as $P = 62{,}008$ N, which is in good correlation with the load from the test, $P = 70{,}460$ N.

To further predict the response of the specimen, the following equation is employed to calculate the maximum mid-span deflections for four-point bending (Davalos et al. 2001):

$$\delta = \frac{23PL^3}{1296D} + \frac{PL}{6(\kappa G_{xz} bd)} \tag{3.74}$$

where κ is the shear correction factor and is approximately 1.0 for this study, G_{xz} is the equivalent core shear stiffness, P is the applied load, L is the span length, and the bending stiffness D is defined as

$$D = b\left[\frac{(d-t_f)^2 t_f}{2} E_f + \frac{(d-2t_f)^3}{12} E_c\right] \tag{3.75}$$

and E_f, E_c, b, d, and t_f denote, respectively, facesheet bending stiffness (computer aided design environment for composites [CADEC]; Barbero 1999), equivalent core bending stiffness (Davalos et al. 2001), beam width, beam depth, and face thickness, as listed in Table 3.18.

Substituting all the values into (3.75), we can obtain

$$\delta = 7.50 \times 10^{-5} P \tag{3.76}$$

The result is illustrated in Figure 3.56, up to $P_{max} = 62{,}008$ N.

3.4.7.2 Delamination

Based on the discussion in Section 3.4.2.5.2, the design equation proposed in Section 3.4.5 is adopted to predict the onset of the delamination. The flat-wise tension test (FWT) (ASTM C297-94) is a standard method to measure

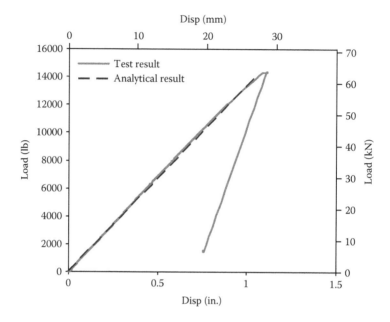

FIGURE 3.56
Load-displacement curve.

the interfacial tensile strength of honeycomb sandwich structures. A series of FWT tests were carried out by Wang (2004), and the test results are summarized in Table 3.19.

It is expected that the specimen with more bonding layers should result in higher interfacial tensile strength, while this is not the case, as observed from Table 3.19. They are somehow randomly distributed, probably due to the variance in the manufacturing process. However, it can be reasonably concluded that the interfacial tensile strength falls within the range of 3.45–6.89 MPa.

TABLE 3.19

Interfacial Tensile Strength

Specimen Type	Interfacial Tensile Stress (MPa)
B1C2	6.26
B2C1	4.62
B2C2	4.92
B2C3	4.03
B3C1	5.90
B3C2	5.78

Source: From Wang, W., Cohesive Zone Model for Facesheet-Core Interface Delamination in Honeycomb FRP Sandwich Panels, PhD dissertation, Department of Civil and Environmental Engineering, West Virginia University, Morgantown, 2004. With permission.

TABLE 3.20

Nominal Interfacial Tensile Strength

	B1C2	B2C2	B3C2	B2C1	B2C3	B3C1	B3C3
Nominal interfacial tensile strength	8.3	12.1	15.1	8.8	11.8	11.8	20.7

Substituting these lower and upper bounds of interfacial tensile strength into (3.66) and (3.71), and using the proposed method as described in Section 3.4.5, we can get the lowest and highest failure loads for the specimen under the four-point bending test, as described in Section 3.4.7, to be 6.89 and 13.78 kN. When comparing with the test data, with a lowest value of 16.68 kN and a highest value of 41.41 kN, we note that the safety factor is from 2.4 to 3.0. Therefore, the method presented in Section 3.4.5 provides a lower bound of the failure load. The reason for this difference is as explained in Section 3.4.5. Since the results are too conservative for design, we may, however, predict the nominal interfacial tensile strength based on the four-point bending test, as shown in Table 3.20.

3.4.8 FE Simulation

As concluded from the experimental results, debonding is a typical failure mode for specimens under four-point bending, where the concept of fracture mechanics should be used for FE modeling. Wang (2004) successfully developed a user-defined element using the cohesive zone model (CZM) and applied it to the four-point bending test. To the best of the authors' knowledge, this is the only work done for the analysis of HFRP sandwich structures, and it is listed here for completeness.

FE modeling of the four-point bending test is performed applying CZM with the mixed-mode linear-exponential constitutive law. The interfacial properties for the cohesive interface element, as listed in Table 3.21, are based on previous experimental measurements. Without experimental data for fracture toughness of modes II and III, it is assumed that $G_{cII} = G_{cIII} = 3\,G_{cI}$.

A 3D finite element model is formulated with ABAQUS. Due to symmetry, only half of the sandwich beam is modeled. The facesheets are modeled with shell elements, and the core is modeled entirely with solid elements. Material degradation within the facesheet-core interfaces during delamination propagation is modeled by embedding cohesive interface elements between the facesheet shell elements and core solid elements.

TABLE 3.21

Fracture Toughness and Interfacial Strength for the Four-Point Bending Test

G_{cI}	$G_{cII} = G_{cIII}$	σ_{c3}	$\sigma_{c1} = \sigma_{c2}$
4.38 N/mm	13.13 N/mm	5.52 MPa	10.34 MPa

With resorting to CZMs, crack initiation and growth could be successfully predicted. As shown in Figure 3.57, the delaminated region is found to be located in the shear loading section of the beam, which is consistent with the observation in the experiments. In Figure 3.58, the finite element result of mid-span deflection vs. applied load is compared to experimental data of the four-point bending test. We can observe that the failure load due to facesheet delamination is accurately predicted. In the numerical simulation,

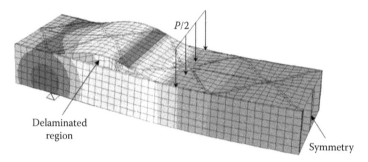

FIGURE 3.57
Finite element model of the four-point bending test of an HFRP sandwich panel with sinusoidal wave core configuration. (From Wang, W., Cohesive Zone Model for Facesheet-Core Interface Delamination in Honeycomb FRP Sandwich Panels, PhD dissertation, Department of Civil and Environmental Engineering, West Virginia University, Morgantown, 2004. With permission.)

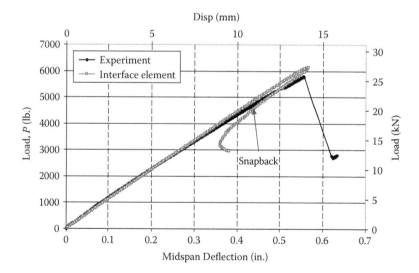

FIGURE 3.58
Finite element results compared to experimental data of the four-point bending test. (From Wang, W., Cohesive Zone Model for Facesheet-Core Interface Delamination in Honeycomb FRP Sandwich Panels, PhD dissertation, Department of Civil and Environmental Engineering, West Virginia University, Morgantown, 2004. With permission.)

severe snapback is induced right after delamination initiation, which could not be captured in the experiment when delamination propagated very quickly, leading to catastrophic sudden collapse of the specimen. Because of the lack of more sophisticated numerical solution methods, the finite element analysis was terminated prematurely, since the global response was successfully captured.

3.4.9 Conclusions

A combined analytical and experimental study of an FRP sandwich panel under out-of-plane shear is presented in this chapter. Analytical models are provided that include delamination considering skin effect, shear crushing, and shear buckling.

Two factors are addressed that contribute to the skin effect: shear warping and bending warping. A closed-form solution, based on proper description of displacement field at the interface, is derived considering shear warping. The accuracy of this method is verified by FE results. The FE model is then applied to study the bending warping effect. The stiffness and the stress distribution subject to skin effect are presented. Critical structural sections are identified, and suggestions for future design considerations are given. Major findings are summarized in Section 3.4.2.

The Rayleigh–Ritz method is employed to study the shear buckling of core panels with two sides elastically restrained. Based on the analytical models, design equations are provided considering delamination, shear crushing of the core, and shear buckling.

Four-point bending tests are carried out according to ASTM standards to study shear strength and shear stiffness of the core materials. In particular, the influence of the facesheet-core interface bonding effect is examined by varying the bonding layers of the specimen. Two cases of bending tests are carried out: longitudinal bending test and transverse bending test. Distinct failure modes were observed for different types of specimens. Design equations are used to predict the failure load due to different failure modes, and good correlations are obtained with experimental results.

3.5 Facesheet Study

3.5.1 Introduction

This section is devoted to studying the strength properties of the facesheet and developing an optimized facesheet configuration (Chen and Davalos 2012). A progressive failure model is developed using the FE method to predict the behavior of laminated composite plates up to failure, where the failure criteria are introduced through prescribed user-defined subroutines. The accuracy of the model is verified through correlations between FE results and existing experimental data. This model is then applied to carry out a parametric study on facesheet. Three variables are included: material properties, including bidirectional stitched fabrics, unidirectional layer of fiber roving, and chopped strand mat; layer thickness; and layer sequences. The quality of each alternative is evaluated based on stiffness and strength performance. In order to further investigate the behavior of facesheet experimentally, coupon samples on selected configurations to evaluate compressive and bending strengths were tested in accordance with ASTM standards. The strength properties in both the longitudinal and transverse directions were evaluated. The dimensions of the coupon specimens vary for different types of tests. The test results are also used to validate the progressive failure model developed in this study. Through this combined experimental and analytical study, the strength properties of facesheet are obtained, which permit the optimization of facesheet design.

3.5.2 Progressive Failure Model

3.5.2.1 Failure Criteria

Various failure criteria for isotropic or composite materials have been proposed. A review of failure criteria of fibrous composite materials was given in Section 3.2. In general, the failure criteria are categorized into two groups: independent and polynomial failure criteria. The maximum stress and strain criteria belong to the first category; they are simple to apply and can define the mode of failure, but they neglect the stress interaction. An interactive criterion such as Tsai-Wu, Hoffman, or Hill includes stress interaction in the failure mechanism, but it does not tell the mode of failure, and it requires some efforts to determine parameters such as F_{12} in Tsai-Wu criterion. Among others, Hashin (1980) provided a three-dimensional failure criterion, which includes fiber tension, fiber compression, matrix tension, and matrix compression. This criterion not only considers the stress interaction, but provides the failure mode, and is therefore adopted in this study. However, Hashin (1980) did not specify the delamination criterion, which becomes significant when the laminate fails due to interlaminar shear failure, as will be shown in Section 3.5.3. Lee (1982) proposed a delamination mode in his 3D analysis, and it is adopted in this study as an addition to Hashin's failure criterion, as shown below.

For a plane stress problem, when considering the transverse shear components, the failure criteria take the following forms (Hashin 1980; Lee 1982):

Tensile fiber mode:

$$\left(\frac{\sigma_{11}}{X_T}\right)^2 + \left(\frac{\sigma_{12}^2 + \sigma_{13}^2}{S_{12}^2}\right)^2 = 1 \quad \sigma_{11} > 0 \tag{3.77}$$

Compressive fiber mode:

$$\sigma_{11} = X_C \quad \sigma_{11} < 0 \tag{3.78}$$

Tensile matrix mode:

$$(\frac{\sigma_{22}}{Y_T})^2 + \frac{\sigma_{23}^2}{S_{23}^2} + (\frac{\sigma_{12}^2 + \sigma_{13}^2}{S_{12}^2}) = 1 \quad \sigma_{22} > 0 \tag{3.79}$$

Compressive matrix mode:

$$\left[(\frac{X_C}{2S_{23}})^2 - 1\right]\frac{\sigma_{22}}{Y_C} + (\frac{\sigma_{22}}{2S_{23}})^2 + \frac{\sigma_{23}^2}{S_{23}^2} + \frac{\sigma_{12}^2 + \sigma_{13}^2}{S_{12}^2} = 1 \quad \sigma_{22} < 0 \tag{3.80}$$

Transverse shear mode:

$$\frac{\sigma_{13}^2 + \sigma_{23}^2}{S_{DS}^2} = 1 \tag{3.81}$$

where σ_{ij} = stress tensor, X_T = tensile failure stress in fiber direction, X_C = compressive failure stress in fiber direction, Y_T = tensile failure stress transverse to fiber direction, Y_C = compressive failure stress transverse to fiber direction, S_{12} = axial failure shear, S_{23} = transverse failure shear, and S_{DS} = interlaminar failure shear.

The material state corresponding to each type of damage is listed in Table 3.22.

TABLE 3.22

Material State

Material State	Elastic Properties					
	E_x	E_y	ν_{xy}	G_{xy}	G_{xz}	G_{yz}
No failure	E_x	E_y	ν_{xy}	G_{xy}	G_{xz}	G_{yz}
Matrix failure	E_x	0	0	0	0	0
Fiber failure	0	0	0	0	0	0
Matrix/fiber failure	0	0	0	0	0	0
Transverse shear damage	E_x	E_y	ν_{xy}	G_{xy}	0	0
Matrix failure/shear damage	E_x	0	0	0	0	0
Fiber failure/shear damage	0	0	0	0	0	0
All damage modes	0	0	0	0	0	0

3.5.2.2 Progressive Failure Analysis

Significant research has been conducted on this topic, as described in the literature review above, in Section 3.2. The objective of this section is to develop a model that uses 2D elements and can still predict the delamination failure, as in the case of a 3D model. Since σ_{33} is negligible considering the thickness-to-length ratio for each layer, only σ_{13} and σ_{23} are considered for delamination, using ABAQUS (2002). A user-defined subroutine is first employed to retrieve the transverse shear stresses from the result file in ABAQUS. Combining with another subroutine to implement the failure criterion displayed in the previous section, the progressive failure analysis is carried out. Due to the nonlinearity after the first-ply failure, displacement control is adopted with the following algorithm.

Obtain stresses for each material point from the previous increment, and retrieve the transverse shear components from the result file:

1. Use Hashin's failure criterion to calculate failure index.
2. Update the field variable according to Table 3.22.
3. Increase the displacement by a given time step.
4. Repeat steps 1 through 4 until ultimate failure is reached.

3.5.3 Verification Study

Greif and Chapon (1993) conducted three-point bending tests of composite beams made of AS4/3502 graphite-epoxy pre-preg tape; the material properties and strength parameters of the test specimens are listed in Table 3.23. Five different laminate types were tested, with two specimens for each type, and the test beam specifications are given in Table 3.24. The reliability of the results was shown by Kim et al. (1996), and these tests are evaluated herein for verification of the progressive failure model.

First, a convergence study is carried out to define the mesh as 30×6 elements. Shell element (S4) in ABAQUS is employed. The predicted vs. experimental load-displacement diagrams for selected graphite-epoxy laminates are shown in Figure 3.59, where good agreement can be observed, although

TABLE 3.23

Material Properties and Strength Parameters

Material Properties	Elastic Properties
$E_1 = 141.2$ GPa	$X_T = 2.343$ GPa
$E_2 = 11.5$ GPa	$X_c = 1.723$ GPa
$G_{12} = 6.0$ GPa	$Y_T = 0.051$ GPa
$v_{12} = 0.3$	$Y_c = 0.223$ GPa
	$S_{12} = 2.343$ GPa
	$S_{DS} = 0.011$ GPa (assumed)

TABLE 3.24

Beam Specifications

Laminate	Lay-Up	No. of Plies	Length (mm)	Width (mm)	Thickness (mm)
A1	$[90_8/0_8]_s$	32	139.7	25.84	4.468
A2	$[90_8/0_8]_s$	32	152.4	25.65	4.547
B1	$[0_8/90_8]_s$	32	127.0	24.13	4.597
B2	$[0_8/90_8]_s$	32	152.4	24.69	4.674
C1	$[(0/90)_8]_s$	32	152.4	25.65	4.470
C2	$[(0/90)_8]_s$	32	152.4	24.33	4.470
D1	$[(45/0/{-}45)_5]_s$	30	152.4	24.26	4.166
D2	$[(45/0/{-}45)_5]_s$	30	152.4	24.26	4.166
E1	$[(0/45/0/{-}45)_3/90/0/0_{1/2}]_s$	29	152.4	24.49	4.039
E2	$[(0/45/0/{-}45)_3/90/0/0_{1/2}]_s$	29	152.4	25.30	4.039

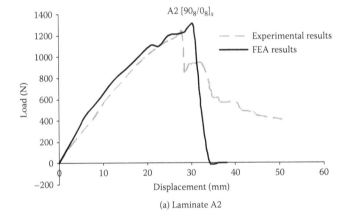

(a) Laminate A2

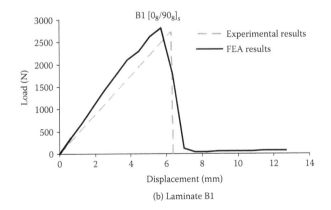

(b) Laminate B1

FIGURE 3.59
Load-displacement paths.

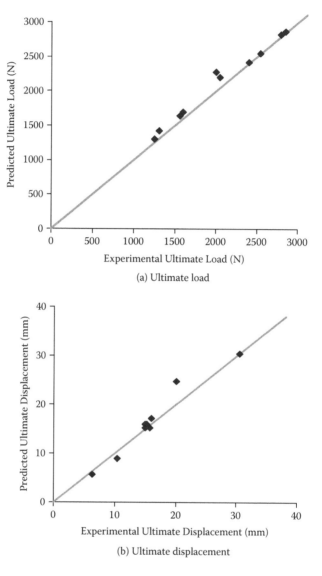

FIGURE 3.60
Comparison of ultimate load and displacement.

some discrepancies for post-failure paths can be noticed. Figure 3.60 compares the ultimate loads and ultimate displacements from FE prediction and test results, illustrating relatively good correlations.

It is worth pointing out that as concluded by Greif and Chapon (1993), beam type B failed due to delamination, which can be easily characterized through a free edge analysis. The other types of laminates followed approximately a progressive failure behavior. As interlaminar shear strength, S_{DS}, is not available in the literature, a value of 0.011 GPa is assumed, as shown in Table 3.23. From the FE analysis, a higher shear stress results at the 0°/90° interfaces. The failure load of type B is highly dependent on the value of S_{DS}, while other types of laminates are not affected too much, which corroborates the relative accuracy of the FE model developed in this study.

3.5.4 Parametric Study on Facesheet

In a sandwich panel, the two stiff facesheets carry the membrane force and the lightweight core resists the out-of-plane shear. As composite action is not considered for an FRP sandwich bridge deck panel, the top and bottom facesheets are respectively subjected to compression and tension of approximate equal magnitude, or vice versa, depending on whether the panel is in the positive or negative bending region. Therefore, the in-plane force is the major concern when designing a facesheet. Since the facesheet can be characterized by longitudinal and transverse directions, four conditions should be considered: tension along the longitudinal direction (TL), compression along the longitudinal direction (CL), tension along the transverse direction (TT), and compression along the transverse direction (CT).

Three variables are included in the parametric study: material properties, including bidirectional stitched fabrics, unidirectional layer of fiber roving, and chopped strand mat; layer thickness; and layer sequences, as listed in Table 3.25. It is noted that Laminate 7 is the facesheet that was being used in industry prior to this study. The material properties given in Table 3.26 are obtained from a previous study by Davalos et al. (2001). The strength parameters given in Table 3.27 are calculated using CADEC (Barbero 1999). As delamination is not a concern for all laminates, $S_{DS} = S_{12}$ is assumed for all the calculations.

TABLE 3.25

Laminate Configuration

Laminate	#1	#2	#3	#4	#5	#6	#7
	1-Bi[a]	1-Bi	1-Bi	1-Bi	6-Bi	1-Bi	1-Bi
	2-Uni[b]	1-ChSM	2-ChSM	1-ChSM	4-ChSM	6-Uni	8-Uni
	1-Bi	1-Bi	1-Bi	4-Uni		1-Bi	1-Bi
	2-Uni	1-ChSM	2-ChSM	1-ChSM		2-ChSM	2-ChSM
	1-Bi	1-Bi	1-Bi	1-Bi			
	4-ChSM[c]	1-ChSM	4-ChSM	4-ChSM			
		1-Bi					
		4-ChSM					
Thickness (mm)	12.45	13.21	12.95	13.21	13.21	10.92	12.95

[a] Bi: CM 3205.
[b] Uni: UM 1810.
[c] 1-ChSM: nominal weight = 457.7 g/m².

TABLE 3.26

Material Properties

Type	E_1 (GPa)	E_2 (GPa)	G_{12} (GPa)	G_{23} (GPa)	ν_{12}	ν_{23}
CM 3205 0°/90°	27.75	8.00	3.08	2.88	0.295	0.39
CM 3205 CSM	11.79	11.79	4.21	2.36	0.402	0.4
UM 1810 0°	30.06	8.55	3.30	3.08	0.293	0.386
UM 1810 CSM	15.93	15.93	5.65	2.96	0.409	0.388
Bond layer ChSM	9.72	9.72	3.50	2.12	0.394	0.401

TABLE 3.27

Strength Parameters (MPa)

Type	X_T	X_C	Y_T	Y_C	S_{12}	S_{23}
CM 3205 0°/90°	1341	404	46	66	46	46
CM 3205 CSM	152	152	152	152	76	83
UM 1810 0°	1452	409	46	65	46	46
UM 1810 CSM	159	159	159	159	79	83
Bond layer ChSM	147	147	147	147	73	83

Either tensile or compressive loads, acting along either longitudinal or transverse directions, are applied to simulated specimens of 20.32×5.08 cm of laminates with different configurations. Typical curves for a balanced laminate (#3) and an unbalanced laminate (#7) are respectively shown in Figures 3.61 and 3.62, where we can see that the compressive load is more critical for both cases. Apparently Laminate 7, which was being used by industry, is not optimized, as the tensile strength along the longitudinal direction is much higher than the compressive strength, whereas the compressive load controls the final design.

Since the axial load is mainly carried out along the longitudinal direction, the CL case is further considered for all configurations. Load-displacement curves are illustrated in Figure 3.63. CL strength for #7 is 219.76 MPa, and the normalized strength based on #7 is shown in Figure 3.64. To illustrate the change of stiffness for each laminate, a static analysis is carried out for a patch load of 88.96 kN acting at the center of a 2.44×2.44 m sandwich panel with a 20.32 cm thick core. The deflection of #7 is 2.54 mm, and the normalized defection based on #7 is shown in Figure 3.65.

From the analysis above, it is shown that when ChSM is introduced into the facesheet, the strength is not affected much, while the stiffness reduces a lot. Consider #3 as an example, the strength is 9% lower and the deflection is 36% higher than those of #7. However, the deflection for #3, which is $L/700$, where L is the span of the deck, is still in the acceptable range.

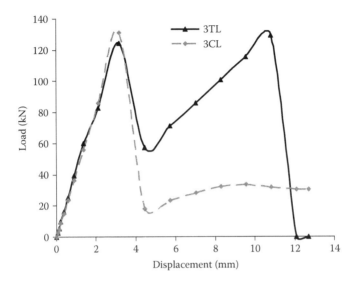

FIGURE 3.61
Load-displacement curve for #3.

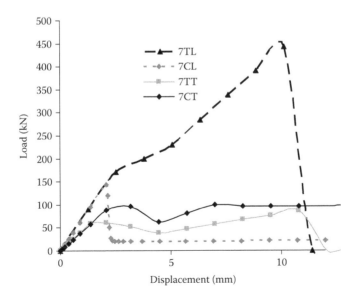

FIGURE 3.62
Load-displacement curve for #7.

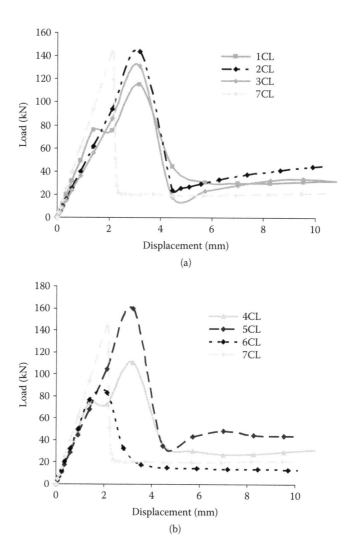

FIGURE 3.63
Load-displacement curves for CL.

3.5.5 Experimental Investigation

Based on the results from the parametric study, three configurations are selected to further study the strength behavior of facesheet, as shown in Table 3.28. Three-point bending tests and compression tests are carried out. Since Laminate 1 is not balanced, the tests are carried out along both longitudinal and transverse directions, resulting in four different types, labeled 1L, 1T, 2, and 3. For completeness, shear test results and stiffness for facesheet laminates are provided in Appendices 3.C and 3.D, respectively.

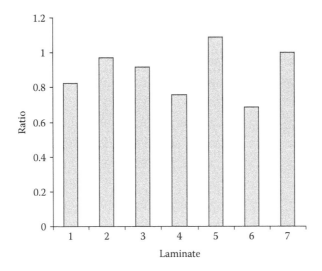

FIGURE 3.64
Normalized failure strength.

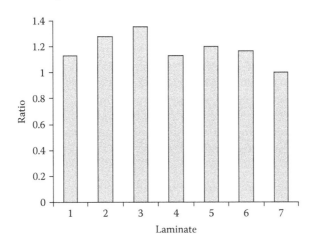

FIGURE 3.65
Normalized deflection.

3.5.5.1 Three-Point Bending Test

3.5.5.1.1 Experimental Setup

The three-point bending test was chosen for the following reasons: (1) the testing apparatus has a simple test setup—no complicated hardware or equipment is required, and (2) the results are relatively easy to interpret. As pointed out by Greif and Chapon (1993), the three-point bending test usually yields good results for material characterization of composites, such as laminate moduli of

TABLE 3.28

Plate Configurations

	Laminate 1(#7)[a] (current)	Laminate 2 (#3)[a]	Laminate 3 (#5)[a]
	2 layers 3.0 oz ChSM	2 layers 3.0 oz ChSM	2 layers 3.0 oz ChSM
	1 layer biaxial	1 layer biaxial	8 layers biaxial
	9 layers uniaxial	1 layer 3.0 oz ChSM	
	1 layer biaxial	1 layer biaxial	
		1 layer 3.0 oz ChSM	
		1 layer biaxial	
		1 layer 3.0 oz ChSM	
		1 layer biaxial	
		1 layer 3.0 oz ChSM	
Thickness	15 mm	16.5 mm	14.2 mm

[a] The number used in parametric study.
Note: Biaxial: CDM 3208
    Uniaxial: CM 1708.
    ChSM: Chopped strand mat.

elasticity, laminate stresses, etc. The test setup is shown in Figure 3.66, which consists of a simply supported beam between two supports with the load applied at the mid-span. The photograph of the setup is shown in Figure 3.67. The dimension of the specimen is 38.1 cm long and 5.08 cm wide. According to ASTM standards (ASTM D790-99), the span is chosen to be 30.48 cm. Four different types were tested, with five specimens for each type. The tests were carried out in a Material Testing Systems, Inc. (MTS) machine. A strain gage was bonded on each specimen at the bottom of the mid-span; the displacement and load were recorded using the internal displacement and load cell transducers. The loading rate was controlled at 8 mm/min.

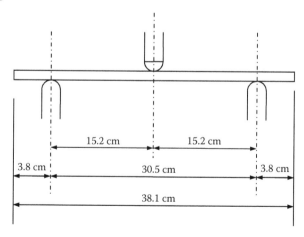

FIGURE 3.66
Three-point bending test setup.

FIGURE 3.67
Photo of test setup.

3.5.5.1.2 Experimental Results

All results are given in terms of applied load vs. displacement or strain at mid-span. For each case, there is good correlation of results, and the failure mechanism observations are reproducible. Therefore, only one plot is shown for each sample. Since there is a ChSM layer at the bottom of the specimen in each configuration, which is the weakest layer, the failure always initiated from this layer and ended by crushing of the top face, as shown in Figure 3.68.

The failure loads are summarized in Table 3.29, where we can see that Laminate 1L is the strongest, followed by Laminate 2, and Laminate 1T has the lowest failure load. It can be seen that when ChSM layers are introduced into the laminate, the strength is not affected much. For Laminates 1L, 2, and 3, the responses showed nearly linear elastic behavior up to failure, as shown in Figure 3.69. At failure initiation, a loud sound of ChSM layer failure was observed with a sudden drop of load. Then the load was redistributed among other layers; the specimen regained some load up to a value that was slightly less than the previous peak load, and then followed by an abrupt failure. For Laminate 1T specimens, after the ChSM layer failed, the sample did not regain any load, which indicated that most

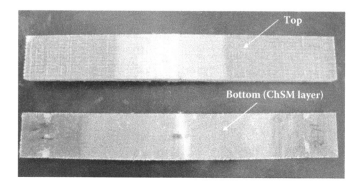

FIGURE 3.68
Failure mode.

TABLE 3.29

Experimental Results for Bending Test

	Laminate 1L	Laminate 1T	Laminate 2	Laminate 3
Failure load (N)	7,251	4,657	7,157	6,561
Standard deviation (N)	467	294	343	249

of the layers failed roughly at the same time. The post-failure path was due to the residual stiffness.

From the above observations, it is concluded that all specimens assume a successive failure mode. The peak load is always associated with the failure of ChSM, which can act as the limiting failure strength of the laminates under bending.

3.5.5.2 Compression Test

Unlike other materials, such as concrete, compressive strength of composite materials is more difficult to measure due to broom-like splitting (termed brooming) at both loaded ends, causing premature failure that is not representative of the actual compressive strength. Therefore, a lot of effort has been devoted to develop appropriate compression test fixtures in order to provide proper boundary conditions. ASTM specifies two methods for compression test: for specimens thinner than 3.175 mm, a support jig is recommended to prevent buckling; and for those thicker than 3.175 mm, the specimen can be tested without any support, which applies to this study. However, the direct application of this method cannot avoid brooming, as will be shown in this section. To solve this problem, Barbero et al. (1999) developed a new fixture. This fixture has been used successfully to determine compressive strength of composites (Makkapati 1994; Tomblin 1994)

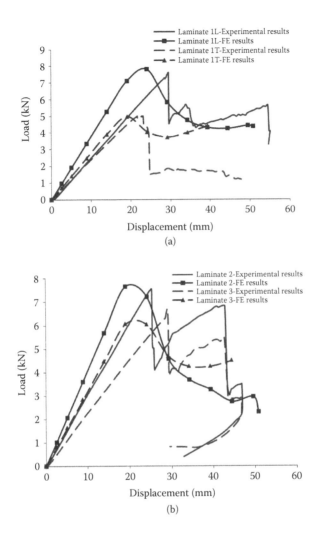

FIGURE 3.69
Load vs. displacement.

and is also adopted herein (see Figures 3.71 and 3.72). The specimens were cut into the dimension of 5.08 × 2.54 cm.

3.5.5.2.1 Experimental Setup

Each half of the compression fixture has two identical 12.7 × 12.7 cm plates, one of which has a rectangular opening in it, so that the specimen can be positioned in the grips using the side support shims that fit in these openings and are at the sides of the specimen. Once the specimen is in position, the specimen's movement is locked by using screws that move the side support shims onto the specimen. The top grip of the fixture can slide on four

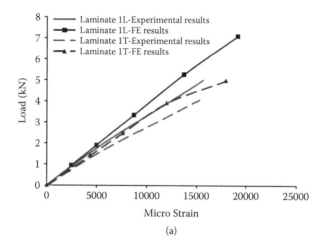

(a)

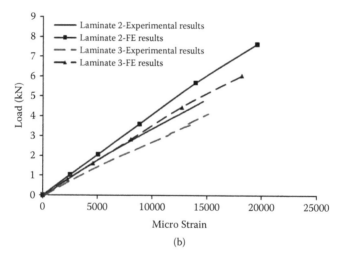

(b)

FIGURE 3.70
Load vs. strain.

guiding posts, which gives a perfect positioning and parallel alignment of the top grip with respect to the bottom grip. Thus, brooming of the ends is avoided by a restraint around the cross section of the sample on the surface of contact with the plate only. A detailed description of the fixture was given by Makkapati (1994).

All specimens were tested in a Baldwin universal testing machine, as shown in Figure 3.73. The fixture uses a cylindrical loading rod, and two rectangular guiding plates, which keep the ends of the specimen intact while loading. When the machine is set for loading, the loading rod pushes the upper half of the compression fixture onto the specimen, to apply the load on the sample. LVDTs were used to measure the movement of the loading block.

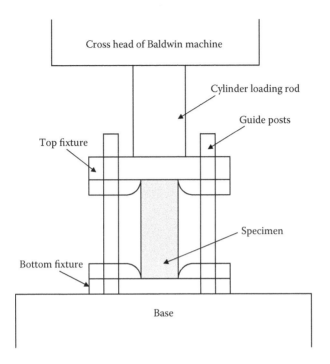

FIGURE 3.71
Experimental setup for compression test.

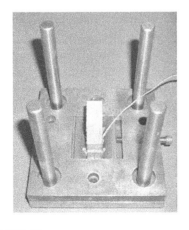 

FIGURE 3.72
Close shot of compression fixture.

FIGURE 3.73
Test setup.

Strain gages were bonded at the mid-height of the specimens to measure the compressive strain.

3.5.5.2.2 Experimental Results

Table 3.30 gives the average failure load and standard deviation for five specimens of each type. It shows that the results obtained from the experimental program are fairly consistent, and the standard deviation is within 10% of the strength of the specimens. As expected, Laminates 1L and 1T correspond to the highest and lowest failure loads, respectively, with Laminates 2 and 3 in between. This corroborates the conclusions drawn from Section 3.5.2, that the strength is not affected much when ChSM is introduced into the face laminates.

During the test, the specimen was intact until the maximum load was reached. It failed with a loud sound and a sudden drop of the load. Figures 3.74 and 3.75 plot the load vs. displacement and load vs. mid-span strain, respectively, showing a typical linear behavior up to failure except for Laminate 3, where some nonlinear behavior can be observed. A typical failure mode is shown in Figure 3.76, indicating that the compressive failure for the laminate was successfully achieved.

TABLE 3.30

Experimental Results for Compression Test

	Laminate 1L	Laminate 1T	Laminate 2	Laminate 3
Failure load (kN)	118.63	61.97	104.05	97.43
Standard deviation (kN)	6.42	5.40	6.84	4.96

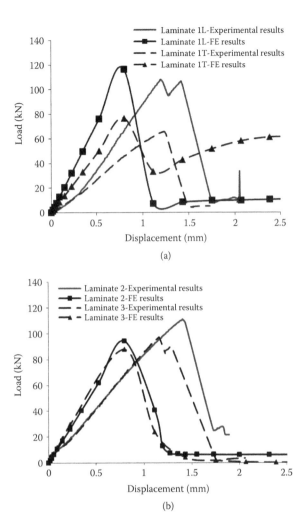

FIGURE 3.74
Load-displacement curve.

As a matter of interest, three specimens were tested according to ASTM standards, i.e., without the end support offered by the compression fixture. The comparison between these two failure modes is shown in Figure 3.77, where we can see an apparent end brooming of the unconstrained specimen and a premature failure by the laminate separation, and as a result, the specimens failed at a much lower load.

3.5.6 Correlation between FE and Experimental Results

The properties of constituent materials used in this study are listed in Table 3.31. The stiffness of properties of composite materials depends on the

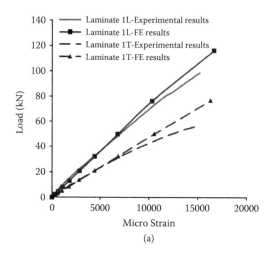

(a)

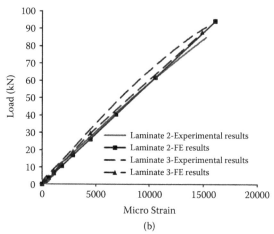

(b)

FIGURE 3.75
Load-strain curve.

FIGURE 3.76
Failed specimen.

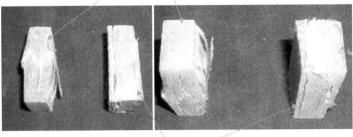

Restrained specimen

Unrestrained specimen

(a) Side View   (b) Top View

FIGURE 3.77
Failure mode comparison.

TABLE 3.31

Properties of Constituent Materials

Material	E (GPa)	G (GPa)	ν	ρ, g/cm³
E-glass fiber	72.4	29.65	0.22	2.56
Polyester (isophthalic) resin	3.65	1.32	0.38	1.1

relative volume of fiber (V_f) and matrix used. For a fiber mat with nominal weight (ω), V_f can be determined from

$$V_f = \frac{\omega}{\rho \cdot t} \tag{3.82}$$

where t is the thickness of the layer and ρ is the density of E-glass fibers. For the face laminates considered, the fiber volume fraction for each layer is computed from (3.82) and shown in Table 3.32. The stiffness of each ply can be predicted from micromechanics models (Luciano and Barbero 1994) and

TABLE 3.32

Layer Properties of Face Laminates

Ply Name	Ply Type	Nominal Weight (g/m²)	Thickness (mm)	V_f
CDM 3208	0°	531	0.49	0.4241
	90°	601.1	0.55	0.4251
	ChSM	256.3	0.25	0.3962
CM 1708	0°	521.8	0.69	0.2947
	ChSM	256.3	0.25	0.3962
Bonding layer	ChSM	915.5	1.91	0.1877

TABLE 3.33

Material Properties

Type	E_1 (GPa)	E_2 (GPa)	G_{12} (GPa)	G_{23} (GPa)	ν_{12}	ν_{23}
CDM 3208 0°/90°	35.89	11.10	3.32	3.03	0.305	0.599
CDM 3208 CSM	17.42	17.42	6.20	6.20	0.406	0.406
CM 1708 0°	23.91	7.39	2.30	2.16	0.333	0.599
CM 1708 CSM	17.42	17.42	6.20	6.20	0.406	0.406
Bonding layer	9.82	9.82	3.51	3.51	0.397	0.397

TABLE 3.34

Strength Parameters (MPa)

Type	X_T	X_C	Y_T	Y_C	S_{12}	S_{23}
CM 3205 0°/90°	1,564	556	51	68	44	44
CM 3205 ChSM	279	279	279	279	139	139
UM 1810 0°	1140	342	50	69	44	44
UM 1810 ChSM	279	279	279	279	139	139
Bond layer ChSM	157	157	157	157	79	79

is listed in Table 3.33. The strength parameters shown in Table 3.34 are calculated using CADEC (Barbero 1999).

3.5.6.1 Three-Point Bending

Using the progressive failure model developed in this chapter, predictions of the compressive strength may be determined, as shown in Table 3.35, from which we can see that predictions from the FE model closely approximate the experimentally obtained results, with a maximum difference of 7.7%. Figure 3.78 compares maximum loads from FE predictions and test results, illustrating a good correlation.

The predicted vs. experimental load-displacement and load-strain curves are shown in Figures 3.69 and 3.70, where good agreement can be observed, although some discrepancies for post-failure paths can be noticed. It is noted that the load-strain curve correlates better than the load-displacement curve, due to the fact that the displacement recorded is the movement

TABLE 3.35

Comparison of Failure Load for Three-Point Bending Test

	Laminate 1L	Laminate 1T	Laminate 2	Laminate 3
Test result (N)	7,255	4,657	7,157	6,561
FE result (N)	7,789	4,969	7,655	6,050
Difference (%)	7.4	6.7	6.9	7.8

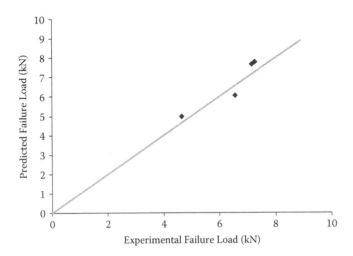

FIGURE 3.78
Comparison of failure load.

of the loading head, and therefore, it cannot represent the actual deflection of the specimen.

3.5.6.2 Compression Test

Following the same approach as described for three-point bending, predictions of the compressive strength are listed in Table 3.36 together with those from tests. Once again, predictions from the FE model closely approximate the experimental results obtained except for Laminate 1T, with a difference of 24.0%. Two factors may contribute to this difference: (1) some nonlinearity was observed during the test, and (2) the compressive strength along the transverse direction is very difficult to predict for a lamina. Figure 3.79 compares maximum loads from FE predictions and test results, illustrating a good correlation.

The predicted vs. experimental load-displacement and load-strain diagrams for selected Laminate 3 are shown in Figures 3.74 and 3.75, where good agreement can be observed.

TABLE 3.36

Comparison of Failure Load for Compression Test

	Laminate 1L	Laminate 1T	Laminate 2	Laminate 3
Test result (kN)	118.63	61.97	104.05	97.43
FE result (kN)	116.40	76.84	94.36	87.73
Difference (%)	1.9	24.0	9.3	9.9

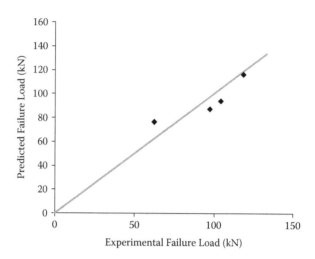

FIGURE 3.79
Comparison of failure load.

TABLE 3.37

Deflection under Patch Load for 2.44 × 2.44 m Plate

Laminate	1L	1T	2	3
Deflection (mm)	2.31	2.98	2.54	2.39

3.5.7 Discussions

To illustrate the change of stiffness for each laminate shown in Section 3.4.5, using the updated material properties, the same static analysis as described in Section 3.4.4 is carried out for a patch load of 88.96 kN acting at the center of a 2.44 × 2.44 m sandwich panel with a 20.32 cm thick core. The results are shown in Table 3.37; the normalized deflection based on Laminate 1L is shown in Figure 3.80. Using the compression test data, the normalized compressive strength based on Laminate 1L is shown in Figure 3.81. The stiffness and strength comparisons provided in Figures 3.80 and 3.81 can be used for design purpose.

3.5.8 Conclusions

A progressive failure model is developed using the FE method to predict the behavior of laminated composite plates up to failure. A parametric study is carried out on strength properties of the facesheet for a sandwich panel using this model. Compressive and bending tests are carried out on selected configurations. From this study, the following conclusions can be drawn:

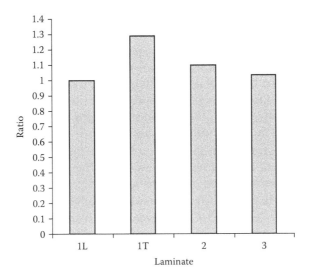

FIGURE 3.80
Normalized deflection under patch load.

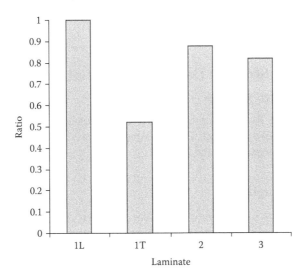

FIGURE 3.81
Normalized strength.

- The progressive failure model developed in this section can be successfully used to predict the behavior of laminated composite plates, as illustrated by the close correlation between FE results and existing experimental data. It is much more efficient than full 3D models and offers great potential for economic parametric studies.

- Interlaminar shear strength should be carefully considered when the delamination occurs prior to other failure modes.
- From the parametric study, it is shown that, for a composite laminate, the compressive strength is lower than the tensile strength. If it is used for the facesheet of a sandwich bridge deck panel, as the top and bottom facesheet are in compression or tension with approximate equal magnitude, the compressive strength of the facesheet is more critical and controls the design.
- If chopped strand mat layer is introduced into the facesheet, the strength is not affected much, while the stiffness is reduced, resulting in a larger deflection under the same loading condition.
- Three-point bending tests were conducted where a progressive failure mode was observed. Compression tests were carried out using a new fixture, where the end brooming is avoided and the true compressive strength is obtained. The results from the bending and compression tests on selected configurations further validate the progressive failure model derived.
- Results from this study can be used for design of facesheets.

Appendix 3.A: Strength Data of Core Materials

3.A.1 Compressive Strength

The same method as described in Section 3.5 is used to carry out the compression test on core material. The results are shown in Table A.1. A typical failure mode is given in Figure A.1.

3.A.2 Shear Strength

A V-notched test is used to find the shear strength. The results are given in Table A.2. A typical shear failure mode is illustrated in Figure A.2, and the test setup is shown in Figure A.3.

TABLE A.1

Compressive Strength

Specimen	Ultimate Load (N)	Thickness (mm)	Width (mm)	Compressive Strength (MPa)
1	27,899	7.125	26.289	148.95
2	26,841	7.074	26.365	143.91
3	28,882	7.455	26.314	147.23
4	30,915	0.762	26.276	154.40
5	27,899	7.099	26.340	145.82

Note: Average strength = 148.06 MPa, standard deviation of strength = 3.99 MPa.

FIGURE A.1
Failed specimen under compression.

TABLE A.2

Shear Strength

Specimen	Ultimate Load (N)	Thickness (mm)	Width (mm)	Shear Strength (MPa)
1	5769	6.833	11.735	71.95
2	5609	7.074	11.659	68.02
3	6405	7.391	11.709	74.01
4	5774	7.366	11.621	67.45
5	6081	7.226	11.760	71.55

Note: Average strength = 70.60 MPa, standard deviation of strength = 2.79 MPa.

FIGURE A.2
Failed specimen under shear.

FIGURE A.3
Test setup for V-notched test.

Appendix 3.B: Derivation of Equilibrium Equation

Consider the element ABCD in Figure B.1, which is cut from the unit cell subject to a shear strain γ. The equilibrium equations for the stresses acting on the $\xi\eta$ plane in the absence of body forces are

$$\frac{\partial \sigma_\xi}{\partial \xi} + \frac{\partial \tau_{\xi\eta}}{\partial \eta} = 0 \qquad (B.1a)$$

$$\frac{\partial \tau_{\xi\eta}}{\partial \xi} + \frac{\partial \sigma_\eta}{\partial \eta} = 0 \qquad (B.1b)$$

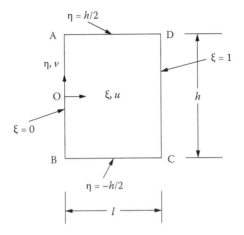

FIGURE B.1
Model cut from the structure.

The stress–strain relationships are

$$
\left\{
\begin{array}{c}
\sigma_\xi \\
\sigma_\eta \\
\tau_{\xi\eta}
\end{array}
\right\}
=
\left[
\begin{array}{ccc}
E' & \nu E' & 0 \\
\nu E' & E' & 0 \\
0 & 0 & G
\end{array}
\right]
\left\{
\begin{array}{c}
\varepsilon_\xi \\
\varepsilon_\eta \\
\gamma_{\xi\eta}
\end{array}
\right\}
\tag{B.2}
$$

where

$$
E' = \frac{E}{1-\nu^2}, \quad G = \frac{E}{2(1+\nu)},
$$

E = Young's modulus, and ν = Poisson's ratio.

The strains can be found through

$$
\varepsilon_\xi = \frac{\partial u}{\partial \xi}
\tag{B.3a}
$$

$$
\varepsilon_\eta = \frac{\partial v}{\partial \eta}
\tag{B.3b}
$$

$$
\gamma_{\xi\eta} = \frac{\partial u}{\partial \eta} + \frac{\partial v}{\partial \xi}
\tag{B.3c}
$$

where u and v are the displacement in the ξ and η directions, respectively. For the consideration of shear warping, we can assume that there is no stretching in the ξ direction. Then we have

$$u = u(\eta) \quad \varepsilon_\xi = 0 \tag{B.4}$$

Equation (B.3) can then be reduced to

$$\sigma_\xi = vE'\varepsilon_\eta \tag{B.5a}$$

$$\sigma_\eta = E'\varepsilon_\eta \tag{B.5b}$$

$$\tau_{\xi\eta} = G\gamma_{\xi\eta} \tag{B.5c}$$

Differentiating (B.5b) and (B.5c) with respect to η and ξ, respectively, substituting into (B.1b), and using (B.3b) and (B.3c), one obtains

$$G\frac{\partial^2 v}{\partial \xi^2} + E'\frac{\partial^2 v}{\partial \eta^2} = 0 \tag{B.6}$$

From the boundary conditions shown in Figure 3.31, $u(\eta) = 0$ at both $\eta = h/2$ and $\eta = -h/2$, and therefore u is negligible throughout the panel. Equation (B.5) becomes

$$\sigma_\xi = vE'(\partial v / \partial \eta) \tag{B.7a}$$

$$\sigma_\eta = E'(\partial v / \partial \eta) \tag{B.7b}$$

$$\tau_{\xi\eta} = G(\partial v / \partial \xi) \tag{B.7c}$$

Appendix 3.C: Shear Test for Facesheet Laminates

3.C.1 Experimental Setup

A shear test (Iosipescu test) was carried out on facesheet laminates for completeness. Due to the challenging efforts and time needed for the specimen preparation, only two types were chosen: Laminate 1L (current configuration) and Laminate 2 (recommended configuration) from Table 3.28, with five specimens each. All specimens were sanded to be around 12.7 mm thick to fit in the fixture. The dimensions of the specimen are given in Figure C.1. The sketch of the test setup is shown in Figure C.2, with two photos given in Figure C.3. All tests were carried out in an MTS machine. Shear strain gage was bonded between the two V-notched sections, and the displacement and

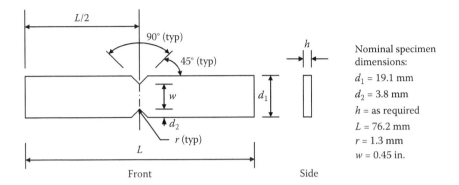

FIGURE C.1
Specimen dimensions.

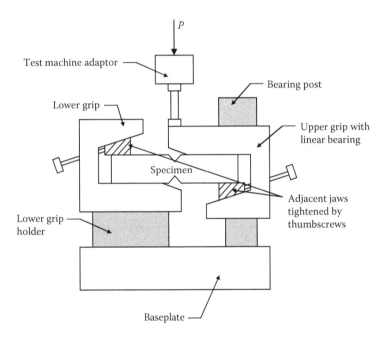

FIGURE C.2
Iosipescu test setup.

load were recorded using the internal displacement and load cell transducers. The loading rate was controlled at 1.27 mm/min.

3.C.2 Experimental Results

All results are given in terms of applied load vs. displacement at mid-span and applied load vs. strain at mid-span, as shown in Figures C.4 and C.5,

FIGURE C.3
Photos of test setup.

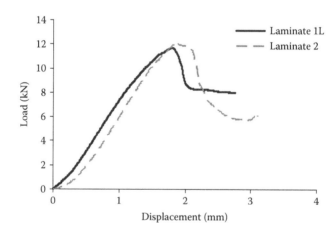

FIGURE C.4
Load-diplacement curves.

which indicates that the specimen followed roughly linear elastic behavior before the failure occurred at the V-notched area. After the specimens failed, the load dropped slowly until the displacement reached the capacity of the testing fixture. Figure C.6 displays several failed specimens, showing a typical shear failure.

The shear strengths for each specimen of Laminates 1L and 2 are given in Tables C.1 and C.2, respectively, from which we can see that when ChSM layers are introduced into the facesheet laminate, the shear strength becomes slightly higher, due to the higher in-plane shear strength provided by random fibers.

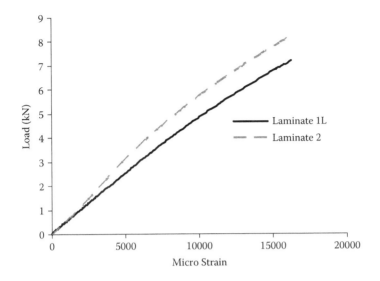

FIGURE C.5
Load-strain curves.

FIGURE C.6
Failure mode.

TABLE C.1

Shear Strength for Laminate 1L

Specimen	Ultimate Load (N)	Thickness (mm)	Width (mm)	Shear Strength (MPa)
1	11,245	12.497	11.125	80.88
2	11,966	12.510	11.328	84.43
3	11,405	12.522	11.43	79.68
4	10,782	12.179	11.176	79.21
5	12,130	12.446	11.303	86.23

Note: Average strength = 82.09 MPa, standard deviation of strength = 3.09 MPa.

TABLE C.2

Shear Strength for Laminate 2

Specimen	Ultimate Load (N)	Thickness (mm)	Width (mm)	Shear Strength (MPa)
1	12,371	12.522	11.138	88.69
2	12,824	12.471	10.731	95.82
3	12,851	12.497	11.379	90.37
4	12,206	12.370	10.998	89.71
5	12,015	12.471	11.316	85.13

Note: Average strength = 89.95 MPa, standard deviation of strength = 3.85 MPa.

Appendix 3.D: Stiffness of Facesheet Laminates and Core Materials

Based on the strain and stress data collected from the compression and shear tests on core materials and facehsheet laminates (Table 3.28), the stiffness can be obtained by fitting the data between 1,000 and 6,000 microstrain, with one example shown in Figure D.1 for determining shear stiffness. The results based on two specimens for each type are listed in Table D.1.

TABLE D.1

Stiffness of Facesheet Laminates and Core Materials

Type	Shear Modulus (MPa)	Compressive Stiffness (MPa)
1L	3,554.9	16,666.7
1T	—	10,835.8
2	4,577.4	13,007.7
3	—	16,002.0
Core materials	3,865.9	9,156.2

References

ABAQUS user's manual. Version 6.3. (2002). HKS, Inc., Providence, RI.

Allen, H.G. (1969). *Analysis and design of structural sandwich panels*. Pergamon Press, Oxford.

Allen, H.G., and Feng, Z. (1998). *Mechanics of sandwich structures*. Kluwer Academic Publisher, Dordrecht, Netherlands, pp. 1–12.

ASTM Designation: C273-00. *Standard test method for shear properties of sandwich core materials*. (2000). ASTM, West Conshohocken, PA.

ASTM Designation: C297-94. *Standard test method for flatwise tensile strength of sandwich constructions*. (2000). ASTM, West Conshohocken, PA.

ASTM Designation: C393-00. *Standard test method for flexure properties of sandwich constructions*. (2000). ASTM, West Conshohocken, PA.

ASTM Designation: D5379-98. *Standard test method for shear properties of composite materials by the V-notched beam method*. (2000). ASTM, West Conshohocken, PA.

ASTM Designation: D790-99. *Standard test method for flexure properties of unreinforced and reinforced plastics and electrical insulating materials*. (2000). ASTM, West Conshohocken, PA.

Barbero, E.J. (1999). *Introduction to composite materials design*. Taylor & Francis, Philadelphia, PA.

Barbero, E.J., Makkapati, S., and Tomblin, J.S. (1999). Experimental determination of compressive strength of pultruded structural shapes. *Composite Science and Technology*, 59, 2047–2054.

Barbero, E.J., and Raftoyiannis, I.G. (1993). Local buckling of FRP beams and columns. *ASCE Journal of Material in Civil Engineering*, 5(3), 339–355.

Blackman, B.R.K., Hadavinia, H., Kinloch, A.J., and Williams, J.G. (2003). The use of a cohesive zone model to study the fracture of fiber composites and adhesively-bonded joints. *International Journal of Fracture Mechanics*, 119, 25–46.

Bleich, F. (1952). *Buckling strength of metal structures*. McGraw-Hill, New York.

Burton, W.S., and Noor, A.K. (1997). Structural analysis of the adhesive bond in a honeycomb core sandwich panel. *Finite Elements in Analysis and Design*, 26, 213–227.

Caprino, G., and Langella, A. (2000). Study of a three-point bending specimen for shear characterisation of sandwich cores. *Journal of Composite Materials*, 34(9), 791–814.

Chen, A., and Davalos, J.F. (2003). Bending strength of honeycomb FRP sandwich beams with sinusoidal core geometry. At Proceedings of the Fourth Canadian-International Composites Conference, CANCOM 2003, Ottawa, Canada, August 19–22.

Chen, A., and Davalos, J.F. (2005). A solution including skin effect for stiffness and stress field of sandwich honeycomb core. Submitted to *International Journal of Solids and Structures*, 42(9), 2711–2739.

Chen, A., and Davalos, J.F. (2007). Transverse shear with skin effect for composite sandwich with honeycomb sinusoidal core. *Journal of Engineering Mechanics ASCE*, 133(3), 247–256.

Chen, A., and Davalos, J.F. (2012). Development of facesheet for honeycomb FRP sandwich panels. *Journal of Composite Materials*, 46(26), 3277–3295.

Christensen, R.M. (1991). *Mechanics of composite materials*. Krieger, Malabar.

Cui, W., and Wisnom, M.R. (1992). A combined stress-based and fracture-mechanics-based model for predicting delamination in composites. *Composites*, 24(6), 467–474.

Cui, W.C., Wisnom, M.R., and Jones, M. (1992). Failure mechanism in three and four point short beam bending tests of unidirectional glass/epoxy. *Journal of Strain Analysis*, 27(4), 235–243.

Cvitkovich, M.K., and Jackson, W.D. (1998). Compression failure mechanisms in composite sandwich structures. At American Helicopter Society 54th Annual Forum, Washington DC, May 20–22.

Davalos, J.F., and Chen, A. (2005). Buckling behavior of honeycomb FRP core with partially restrained loaded edges under out-of-plane compression. *Journal of Composite Materials*, 39(16), 1465–1485.

Davalos, J.F., Qiao, P., Xu, X.F., Robinson, J., and Barth, K.E (2001). Modeling and characterization of fiber-reinforced plastic honeycomb sandwich panels for highway bridge applications. *Composite Structures*, 52, 441–452.

DeTeresa, S.J., Freeman, D.C., Groves, S.E., and Sanchez, R.E. (1999). Failure under multiaxial stresses of component materials for fiber composite sandwich construction. In *Proceedings of Twelfth International Conferences on Composite Materials (ICCM-12)*, Paris, France, July 5–9, 1999, p. 198.

Echaabi, J., Trochu, F., and Gauvin, R. (1996). Review of failure criteria of fibrous composite materials. *Polymer Composites*, 17(6), 786–798.

Elawadly, K.M. (2003). On the interlaminar shear stress response for E-glass/epoxy composite. *Journal of Composite Materials*, 37(23), 2149–2158.

El-Sayed, S., and Sridharan, S. (2002). Cohesive layer model for predicting delamination growth and crack kinking in sandwich structures. *International Journal of Fracture Mechanics*, 117, 63–84.

FEMAP user's manual. Version 8.1. (2001). Enterprise Software Products, Exton, PA.

Greif, R., and Chapon, E. (1993). Investigation of successive failure modes in graphite/epoxy laminated composite beams. *Journal of Reinforced Plastics and Composites*, 12, 602–621.

Halpin, J.C., and Kardos, J.L. (1978). Strength of discontinuous reinforced composites. I. Fiber reinforced composites. *Polymer Engineering and Science*, 18(6), 496–504.

Hann, H.T. (1975). On approximation for strength of random fiber composites. *Journal of Composite Materials*, 9, 316–326.

Hashin, Z. (1980). Failure criteria for unidiretional fiber composites. *Journal of Complied Mechanics*, 47, 329–334.

Hwang, W.C., and Sun C.T. (1989). Failure analysis of laminated composites by using iterative three-dimensional finite element method. *Computers and Structures*, 33(1), 41–47.

Kelsey, S., Gellatly, R.A., and Clark, B.W. (1958). The shear modulus of foil honeycomb cores. *Aircraft Engineering*, 30, 294–302.

Kim, R., and Crasto, Y. (1992). A longitudinal compression test for composites using a sandwich specimen. *Journal of Composite Materials*, 26(13), 1915–1929.

Kim, Y. (1995). A layer-wise theory for linear and failure analysis of laminated composite beams. PhD dissertation, Department of Civil and Environmental Engineering, West Virginia University, Morgantown.

Kim, Y., Davalos, J.F., and Barbero, E.J. (1996). Progressive failure analysis of laminated composite beams. *Journal of Composite Materials*, 30(5), 536–560.

Kollar, L.P. (2002). Buckling of unidirectional loaded composite plates with one free and one rotationally restrained unloaded edge. *Journal of Structural Engineering*, 128(9), 1202–1211.

Kroll, L., and Hufenbach, W. (1997). Physically based failure criterion for dimensioning of thick-walled laminates. *Applied Composite Materials*, 4(5), 321–332.

Kumar, P., Chandrashekhara, K., and Nanni, A. (2003). Testing and evaluation of components for a composite bridge deck. *Journal of Reinforced Plastics and Composites*, 22(5), 441–461.

Lee, H.S., Lee, J.R., and Kim, Y.K. (2002). Mechanical behavior and failure process during compressive and shear deformation of honeycomb composite at elevated temperatures. *Journal of Materials Science*, 379(6), 1265–1272.

Lee, J.D. (1982). Three dimensional finite element analysis of damage accumulation in composite laminate. *Computers and Structures*, 15(3), 335–350.

Lingaiah, K., and Suryanarayana, B.G. (1991). Strength and stiffness of sandwich beams in bending, *Experimental Mechanics*, 3, 1–7.

Lopez-Anido, R., Davalos, J.F., and Barbero, E.J. (1995). Experimental evaluation of stiffness of laminated composite beam elements under flexure. *Journal of Reinforced Plastics and Composites*, 14, 349–361.

Luciano, R., and Barbero, E.J. (1994). Formulas for the stiffness of composites with periodic microstructure. *International Journal of Solids and Structures*, 31(21), pp. 2933–2944.

Makkapati, S. (1994). Compressive strength of pultruded structural shapes. Master's thesis, Department of Mechanical and Aerospace Engineering, West Virginia University, Morgantown.

Mouritz, A.P., and Thomson, R.S. (1999). Compression, flexure and shear properties of a sandwich composite containing defects. *Composite Structures*, 44, 263–278.

Niu, K., and Talreja, R. (1998). Modeling of wrinkling in sandwich panels under compression. *Journal of Engineering Mechanics*, 125(8), 875–883.

Noor, A.K., Burton, W.S., and Bert, C.W. (1996). Computational models for sandwich panels and shells. *Applied Mechanics Reviews, ASME*, 49(3), 155–199.

Ochoa, O.O., and Engblom, J.J. (1987). Analysis of progressive failure in composites. *Composites Science and Technology*, 28, 87–102.

Papadopoulos, L., and Kassapoglou, C. (2004). Shear buckling of rectangular composite plates with two concentric layups. *Journal of Reinforced Plastics and Composites*, 23(1), 5–16.

Qiao, P., Davalos, J.F., and Wang, J. (2001). Local buckling of composite FRP shapes by discrete plate analysis. *Journal of Structural Engineering*, 127(3), 245–255.

Qiao, P., and Zou, G. (2003). Local buckling of composite fiber-reinforced plastic wide-flange sections. *Journal of Structural Engineering*, 129(1), 125–129.

Qiao, P.Z., and Huo, X.P. (2011). Explicit local buckling analysis of rotationally-restrained orthotropic plates under uniform shear. *Composite Structures*, 93(11), 2785–2794.

Reddy, J.N. (1999). *Theory and analysis of elastic plates*, Taylor & Francis, Philadelphia, PA.

Reddy, Y.S., and Reddy, J.N. (1993). Three dimensional finite element progressive analysis of composite laminates under axial extension. *Journal of Composites Technology and Research, JCTRER*, 15(2), 73–87.

Shan, L.Y., and Qiao, P.Z. (2008). Explicit local buckling analysis of rotationally restrained plates under uniaxial compression. *Engineering Structures*, 30(1), 126–140.

Spottswood, S.M., and Palazotto, A.N. (2001). Progressive failure analysis of a composite shell. *Composite Structures*, 53(1), 117–131.

Tan, S.C. (1991). A progressive failure model for composite laminates containing openings. *Journal of Composite Materials*, 25, 556–577.

Tan, S.C., and Perez, J. (1993). Progressive failure of laminated composite with a hole under compressive loading. *Journal of Reinforced Plastics and Composites*, 12, 1043–1057.

Theotokoglou, E.E. (1996). Analytical determination of the ultimate strength of sandwich beams. *Applied Composite Materials*, 3, 345–353.

Thomson, R.S., Shah, K.Z., and Mouritz, A.P. (1989). Shear properties of a sandwich composite containing defects. *Composite Structures*, 11, 101–120.

Timoshenko, S.P., and Gere, J.M. (1961). *Theory of elastic stability*. McGraw-Hill Book Company, New York.

Tolson, S., and Zabaras, N. (1991). Finite element analysis of progressive failure in laminated composite plates. *Computers and Structures*, 38(3), 361–376.

Tomblin, J.S. (1994). Compressive strength models for pultruded glass fiber reinforced composites. PhD dissertation, Department of Mechanical and Aerospace Engineering, West Virginia University, Morgantown.

Triantofillou, T.C., and Gibson, L.J. (1989). Debonding in foam-core sandwich panels. *Materials and Structures*, 22, 64–69.

Tsau, L.-R., and Plunkett, R. (1993). Finite element analysis of progressive failure for laminated FRP plates with inplane loading, *Engineering Fracture Mechanics*, 45(4), 529–546.

Ungsuwarungsri, T., and Knauss, W.K. (1987). The role of damage-soften material behavior in the fracture of composites and adhesives. *International Journal of Fracture Mechanics*, 35, 221–241.

Vinson, J.R. (1999). *The behavior of sandwich structures of isotropic and composite material*. Technomic Publishing Company, Lancaster, PA.

Vinson, J.R., and Sierakowski, R.L. (1987). *The behavior of structures composed of composite materials*. Martinus Nijhoff Publishers, Dordrecht, Netherlands.

Waas, A.M., and Schultheisz, C.R. (1996). Compressive failure of composites. II. Experimental studies. *Progress in Aerospace Sciences*, 32, 43–78.

Wang, W. (2004). Cohesive zone model for facesheet-core interface delamination in honeycomb FRP sandwich panels. PhD dissertation, Department of Civil and Environmental Engineering, West Virginia University, Morgantown.

Xu, X.F., Qiao, P., and Davalos, J.F. (2001). Transverse shear stiffness of composite honeycomb core with general configuration. *Journal of Engineering Mechanics*, 127(11), 1144–1151.

Zenkert, D. (1991). Strength of sandwich beams with interface debondings. *Composite Structures*, 17, 331–350.

Zenkert, D. (1995). *An introduction to sandwich construction*. Chamelon Press, London.

Zhang, J., and Ashby, M.F. (1992). Out-of-plane properties of honeycombs. *International Journal of Mechanical Sciences*, 34(6), 475–489.

4

Mechanical Shear Connector for FRP Decks

4.1 Introduction

There are two major types of connections for fiber-reinforced polymer (FRP) decks to steel girders: adhesive connection, which is formed by applying adhesive glue at the deck-stringer interface; and mechanical connection, which is formed by connecting FRP decks to steel girders using steel clamps, bolts, or shear studs.

A series of experimental studies on adhesive connections were conducted by Keller and Gürtler (2005). Two large-scale T-beams were constructed with pultruded cellular FRP decks and steel girders, and their stiffness, strength, and fatigue responses were investigated by static and cyclic tests. It was shown that composite action could be achieved using the adhesive bond, leading to increases in stiffness and strength of the FRP deck–steel girder system. No stiffness degradation was observed under fatigue loading. It is noted, however, that the quality of the adhesive bond can be affected by environmental exposure to moisture and temperature. Also, it is difficult to achieve proper contact and an effective bond in the field, and to implement quality control measures.

For bolted and clamped connections, the installation process is quite labor-intensive, as the work must be performed from underneath the bridge deck (Righman et al. 2004). Therefore, shear stud connections are more favorable, and their function is conceptually related to studs used for concrete decks. Moon et al. (2002) developed a shear stud-type connection (Figure 4.1) for a trapezoidal sandwich panel, known as Martin Marietta Composites (MMC) Gen4 FRP deck. The connector was designed to resist the shear force at the deck-stringer interface and to develop composite action. The connector consisted of shear studs and enclosures within the deck. The shear studs were prewelded on the steel stringers, with two or three studs in one group. An opening was cut out from the FRP deck to accommodate the studs, and after the FRP deck was in place, the opening was filled with expansive concrete grout. Static tests on push-out specimens showed that for the specific configuration shown in Figure 4.1, this connection could sustain a load of up to 347 kN with 12.7 mm displacement. A substantial inelastic deformation

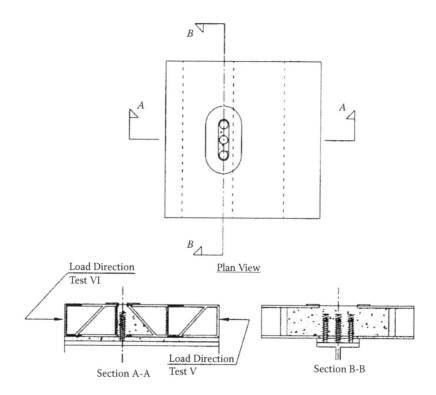

FIGURE 4.1
Schematic of connection. (From Moon, F. A., Eckel, D. A., and Gillespie, J. W., *Journal of Structural Engineering*, 128(6), 762–770, 2002.)

occurred before failure, due mainly to yielding of the shear studs. The specimens were further subjected to a 56 kN fatigue load of up to 10.5 million cycles, which was considered to be equivalent to 75 years of bridge design life span. The specimens did not show any obvious damage throughout the loading history, and the stiffness remained almost constant. These results indicated that the shear connection had adequate fatigue resistance.

In 2004, Keelor et al. conducted a field test on a short-span bridge with an FRP deck in Pennsylvania, using a similar connection as the one shown in Figure 4.1. The bridge had a pultruded FRP deck that was attached to steel stringers using the same conceptual shear connection developed by Moon et al. (2002), and it was designed as fully composite. The span was 12.6 m long, and there were five steel girders equally spaced at 1.8 m. The spacing of the shear connections was 0.6 m, and each connection consisted of two-headed shear studs welded side by side on the top flange of the girders. The field test showed that at service load, this bridge system was able to achieve full composite action. The effective width was close to 90% of the girder spacing for interior girders and approximately 75% of half girder spacing for exterior girders.

The stud-type connection described above can transfer the required shear force and develop composite action for FRP decks, but it requires labor-intensive and careful placement of grout. Since this connection is expected to achieve full composite action, there is a concern with high stress concentration within the enclosed grout area, and potential for cracking and degradation, causing a negative impact on the integrity of the deck and connection.

To address the above concerns, a prototype shear stud sleeve-type connection was developed by Davalos et al. (2000) and further studied by Righman et al. (2004), as shown in Figure 4.2, which reported that this connection can be easily manufactured and can satisfy requirements for structural performance and field assembly and implementation. This connection can prevent lifting of the FRP deck, and it offers versatility of applications to most FRP decks, ease of installation and replacement, and structural efficiency. The connection consists of a threaded shear stud welded to the top flange of the supporting steel girder and housed inside of steel sleeves that are installed within a hole drilled through the FRP deck. It can provide a secure connection, preventing uplift of the bridge deck by way of the top washer of the steel sleeve; i.e., as the nut is tightened, the top washer exerts a downward force on the FRP deck. Another important feature is that this connection has the ability to be used in conjunction with any type of commercially available FRP bridge deck, including pultruded and sandwich decks. The height of the sleeves can be easily adjusted, creating a functional connection for various deck thicknesses. Construction of this connection would involve welding the threaded studs to the girder (similar to what is done in the construction of reinforced concrete bridge decks), placing the FRP deck with the sleeves installed, placing and tightening the nuts on the studs to secure the deck, and then covering the cavities within the sleeves using a granular material (sand) for accessibility, or a polymer binder, or a cap. The installation of this connection is fabricated by allowing all labor to be performed from above the bridge deck. As a result, construction time can be reduced, providing some cost savings. In the event there is a need for the replacement of the deck, the connection can easily be released by removing the nut and washer. An additional attribute of this connection is its structural efficiency. Since composite materials have relatively low compressive and shear strength properties, this connector minimizes these stresses by way of the protective steel sleeves within the relatively large contact surfaces between the oversized hole and surrounding steel tubing. Moreover, steel shear studs have shown favorable fatigue performance characteristics (Slutter and Fisher 1966).

Because of the advantages mentioned above, this connection has been implemented in the field, and more recently applied to the Wildcat Creek Bridge, Indiana (Machado et al. 2008). Following the promising developments of the initial conceptual preliminary design (Davalos et al. 2000; Righman et al. 2004), this chapter presents an experimental study to evaluate stiffness, strength, and fatigue performance characteristics of this connection, to

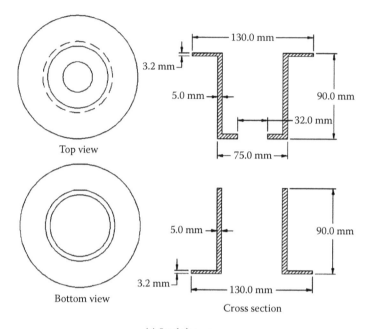

(a) Steel sleeves

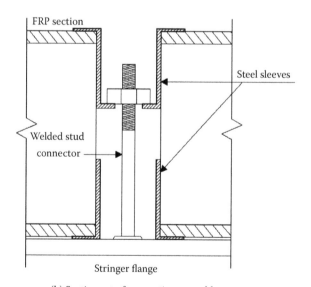

(b) Section cut of connection assembly

FIGURE 4.2
Shear connection details.

develop design criteria based on component level evaluation, and to define the degree of composite action based on testing at the system level.

The three most common major testing methods for shear connection evaluation are push-out test, beam test, and scaled-bridge test. A push-out test specimen consists of a bridge deck section attached to a single shear connector. Beam and scaled-bridge test specimens are considered representative of full-scale bridges, and consist of assemblies of steel beams, bridge decks, and shear connectors. Slutter and Fisher (1966) conducted fatigue tests on 56 push-out specimens, and they used both stud and channel connectors. Mainstone and Menzies (1967) conducted fatigue tests on both push-out specimens and T-beams, and they studied three common types of shear connectors: stud, channel, and bar. Oehlers and Bradford (1995) proposed an alternative test method on beam specimens, where the static strength and fatigue resistance of shear connections were integrated and related; the beam test specimens were subjected to fatigue loading of a predetermined number of load cycles, and then statically loaded to failure. This method was used to recommend a design approach for shear connections. Righman et al. (2004) tested a scaled bridge with an FRP deck that incorporated sleeve-type shear connectors. The bridge was subjected to static load tests, and the resulting reactions and deflections from these tests were compared with finite element models of the system. Recently, Majumdar et al. (2009) constructed a two-bay section (9.45 × 6.7 m) of a bridge with an FRP deck and tested it under different probable loading scenarios to explore its construction feasibility, serviceability, and durability using stud connectors.

To limit testing variables, push-out tests are suitable to study a connection system at a component level, to evaluate typically the static and fatigue strengths of a single connector, where a section of the deck and the attachment are subjected to in-plane shear loading by avoiding in-plane twisting and out-of-plane bending. Both beam-type and scaled-bridge test models are suitable to study a connection at a system level, to evaluate, for example, degree of composite action, load distribution factor, and effective flange width. This chapter is focused on the connection system at a component level, while the connection system at the system level based on scaled-bridge tests will be covered in Chapter 5.

4.2 Prototype Shear Connection

The shear connection consists of two steel sleeves, designated as top and bottom sleeves, as shown in Figure 4.2. The top sleeve is 90 mm long and consists of a 75 mm diameter tubing with two welded washers. The top washer has a 130 mm outside diameter, and the bottom washer has a 32 mm inside diameter. The bottom sleeve is a 90 mm long and 75 mm diameter tubing welded to a bottom washer with a 130 mm outside diameter. The height of the tubing can be varied in order to accommodate FRP panels of different thicknesses. The function of the tubing, or sleeve, is to provide a protective enclosure for the panel and to allow mechanical attachment to the welded shear stud. The top exterior washer serves to clamp the panel to the stringer, while protecting the FRP panel by distributing the stresses over an adequate contact area. The smaller washer inside the tubing, fitted with an additional pressure washer under the nut, is used to secure the sleeves to the welded shear stud. During installation, the steel sleeves are tightly fitted into drilled holes in the FRP deck panel without using any bonding, and then the deck panel is placed over the girders and through the prewelded shear studs. Then the pressure washer and nut are installed for all connectors. The interface shear force is transferred from the FRP panel to the inside washer and tubing through the bearing, and then to the shear stud and steel girder. Details of the interface shear transferring mechanism are discussed in the following section.

4.3 Push-Out Test

4.3.1 Specimen and Test Setup

As shown in Figures 4.3 and 4.4, the push-out specimen consisted of a honeycomb FRP (HFRP) sandwich panel, fitted with a single shear connection at the center (Davalos et al. 2011). The 0.2 m deep panel was manufactured by Kansas Structural Composites, Inc. (KSCI) with an in-plane dimension of 0.9×0.9 m². The sandwich configuration is schematically shown in Figure 1.1. The core was 0.17 m high and each facesheet was 15 mm thick. The push-out specimen was attached to a floor beam connected to the laboratory's strong

Rubber roller

FIGURE 4.3
Shear connection test setup.

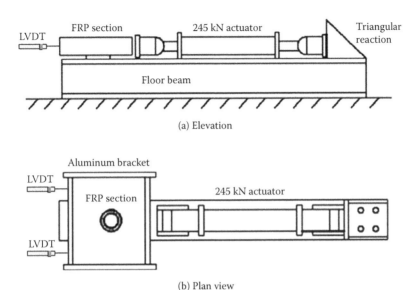

FIGURE 4.4
Schematic shear connection test setup.

floor. The specimen was loaded horizontally, using a 245 kN actuator as shown in Figure 4.4, to simulate the interface shear transfer in composite bridge decks. The vertical position of the actuator was adjusted to minimize eccentricity and avoid out-of-plane bending during the loading. In order to prevent the panel from in-plane rotation and to distribute the horizontal force more evenly, an aluminum frame was installed around the FRP panel and then connected to the actuator head fitted with a swivel to avoid bending. The side channels of the aluminum frame were constrained by hard rubber-covered steel rollers on ball bearings that were attached horizontally to side steel plates to provide lateral support with minimum frictional effects. Two linear variable displacement transducers (LVDTs) were used to record the horizontal displacement of the specimen.

4.3.2 Test Procedures

The tests were conducted in two phases. Phase I included eight specimens, numbered S1 through S8, statically loaded until failure. A preliminary test was first conducted on specimen S1 to study its failure mode and evaluate its damage evolution. The specimen was disassembled at a load increment of 11 kN, and then was reassembled and reloaded to the next load level. Tests on specimens S2 to S8 were conducted continuously using a displacement control at a rate of 3 mm/min up to failure.

In phase II, fatigue tests were conducted on 10 similar specimens numbered F1 through F10. The two control parameters used were stress ranges

TABLE 4.1

Fatigue Test Results

Test	Fatigue Load (kN)		Stress Range (MPa)	Load Ratio	Rate (Hz)	Life Cycles (million)
	Min	Max				
F1	11	47	93	30%	4	2.58
F2	11	29	46	15%	4	13.84
F3	11	35	62	20%	4	8.36
F4	11	35	62	20%	4	10.25
F5	11	59	124	40%	4	1.01
F6	11	59	124	40%	4	1.55
F7	11	83	186	60%	4	0.39
F8	11	83	186	60%	4	0.69
F9	11	95	217	70%	4	0.13
F10	11	95	217	70%	4	0.25

on the shear stud and corresponding fatigue life cycles. A pilot test on specimen F1 was conducted to obtain preliminary data and define subsequent testing protocols, with a load range corresponding to 30% of the connection's ultimate strength. Subsequently, tests on F2 to F10 were conducted at five different loading ranges, with details shown in Table 4.1. All specimens were subjected to cyclic loading with a loading frequency of 4 Hz, which was within the range of fundamental frequency of 2 to 5 Hz for typical highway steel bridges induced by cyclic standard truckloads.

4.3.3 Test Results and Discussions

For static strength and load-displacement formulation (*P-Δ* curve), preliminary tests and inspections on specimen S1 revealed the deformation and failure mechanism of the connection. After initial slip, the force transfer mechanism was bearing between the FRP deck and top sleeve. The top sleeve displayed warping for both outside and inside washers, and continued to deform as loading increased, as shown in Figure 4.5(a) and (b). The effect of friction force between the FRP deck and steel girder was negligible because both the bottom washer and the resin-coated surface of the FRP deck were very smooth. The deformation for the bottom sleeve was first observed at the contact zone between the bottom sleeve and the shear stud, at about 22 kN. And then it continued to deform as the load increased, with a more significant deformation than the top sleeve, as shown in Figure 4.5(c) and (d).

For the FRP deck, there was virtually no damage for the top facesheet. For the bottom facesheet, significant deformation was observed at the location where the FRP made contact with the shear stud and sleeve, as shown in Figure 4.5(e) and (f). The connection's failure mode was a fracture of the root of the shear stud, as shown in Figure 4.6, and delamination of the bottom facesheet, as shown in Figure 4.7.

(a) Top view of top sleeve

(b) Bottom view of top sleeve

(c) Top view of bottom sleeve

(d) Side view of bottom sleeve

(e) Bottom FRP facesheet at
yield of shear stud

(f) Bottom FRP facesheet at failure

FIGURE 4.5
Deformations of components of shear connection.

Specimens S2 through S8 were tested continuously without any interruption, with load-displacement curves shown in Figure 4.8, which displays three load-displacement stages. The stiffness is very low at the first stage up to about 22 kN, and then increases significantly at the second stage after the bottom sleeve makes contact with the shear stud. At the third post-yielding stage, the specimen continues to deform at almost a constant load until failure, displaying a ductile behavior, which comes mainly from the shear stud yielding.

FIGURE 4.6
Fracture of shear studs.

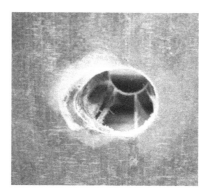

FIGURE 4.7
Delamination of bottom FRP facesheet.

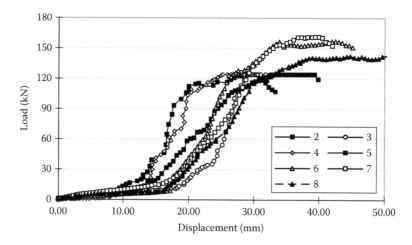

FIGURE 4.8
Load-displacement curves for specimens S2–S8.

TABLE 4.2

Static Strength of Shear Connections

Specimen	Yield Strength (kN)	Ultimate Strength (kN)
S2	112	123
S3	113	126
S4	109	123
S5	103	124
S6	122	153
S7	125	161
S8	115	141
Average	114	136
Standard deviation	7.5	15.9
COV	6.6%	11.7%

The connection yielding and ultimate loads are listed in Table 4.2. Although the curves in Figure 4.8 are scattered due to the initial gap between the steel sleeves and FRP deck, the strength values at the three stages are within a close range. The yielding load varied from 103 to 125 kN, with a coefficient of variation (COV) of 6.6%, and the ultimate load varied from 123 to 161 kN, with a COV of 11.7%. The discrepancies in strength values were mainly due to (1) the manufacturing and material nonuniformity of the facesheets and (2) the manufacturing imperfection of the steel sleeves. To be conservative, the lower bound values were taken as design values. Therefore, the recorded load-displacement curves can be idealized as a segmentally linear model, as shown in Figure 4.9, with three ranges that can be described as

$$\begin{cases} k = 1.5 \text{ kn/mm}, & \Delta = 0\text{–}15 \text{ mm} \\ k = 7.9 \text{ kn/mm}, & \Delta = 15\text{–}25 \text{ mm} \\ k = 1.4 \text{ kn/mm}, & \Delta > 25 \text{ mm} \end{cases}$$

where k is the stiffness of the connection.

To avoid damage to the FRP deck and to simulate field operations commonly used in practice for bolted connection, the torque applied to the shear stud bolt during the shear connection installation was not a variable in this study. Rather, the bolt was only "snug tightened," corresponding to the definition that the installed torque is attained by full effort of an ironworker with an ordinary spud wrench.

For fatigue strength and the *S-N* curve, by inspection of the pilot test specimen F1, the fatigue crack was initiated around the perimeter of the stud shank and weld area. As the load cycles increased, the crack extended into the steel base plate (e.g., steel beam flange), causing a concave depression, and eventually the shear stud was sheared off, as shown in Figure 4.6. Local

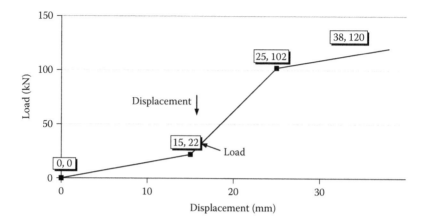

FIGURE 4.9
Segmentally linear load-displacement curve.

crushing of the bottom FRP facesheet was observed at the contact point with the shear stud root, as shown in Figure 4.7, which was similar to the failure mode from static testing.

The fatigue test results are shown in Table 4.1 and Figure 4.10 in terms of fatigue load cycles and stress ranges, which can be fitted by a logarithm function as

$$logN = 7.6 - 0.01096S \tag{4.1}$$

where N is number of load cycles and S is the stress range of the shear connection in MPa.

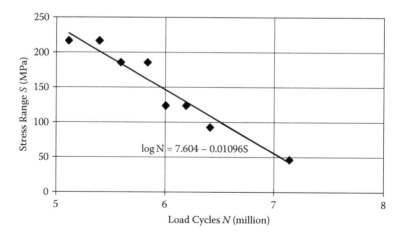

FIGURE 4.10
S-N curve of shear connection.

Moon et al. (2002) reported that a load cycle of 10.5 million corresponded to 75 years of bridge service life span. Based on this criterion, the threshold for the stress range can be calculated as 53 MPa from (4.1). The shear connection detail is defined as category A from AASHTO specifications (2004).

4.4 Conclusions

This chapter presents a detailed experimental study of a mechanical shear connection for FRP bridge decks supported by steel girders at the component level using a push-out test.

Static and fatigue tests were conducted on push-out specimens, which consisted of a honeycomb FRP sandwich deck panel fitted with a single shear connection at the center. The interfacial shear force transfer mechanism was bearing between the deck and shear connector. The P-Δ (load-displacement) curve was approximately simulated as a three-stage segmentally linear model based on results obtained from the static tests. The stiffness was very low at the first stage, and then increased significantly at the second stage after the bottom sleeve made contact with the shear stud. At the third post-yielding stage, the specimen continued to deform at almost a constant load until failure, displaying a ductile behavior. Based on the results from the fatigue test, the S-N (stress range–fatigue life) curve can be fitted by a logarithm function. Both the P-Δ and S-N curves were used to develop design formulas, which can be used to define the spacing of shear connections and the size of the shear stud for given static and fatigue loads.

References

AASHTO. (2004). *AASHTO LRFD bridges design specifications*. 3rd ed. American Association of State Highway and Transportation Officials, Washington, DC.

Davalos, J.F., Barth K.E., and Qiao, P. (2000). *Development of an innovative connector system for fiber reinforced polymer bridge decks to steel girders*. NCHRP Stage I Report. National Cooperative Highway Research Program, Washington, DC.

Davalos, J.F., Chen A., and Zou, B. (2011). Stiffness and strength evaluations of a new shear connection system for FRP bridge decks to steel girders. *Journal of Composites for Construction ASCE*, 15(3), 441–450.

Keelor, D.C., Luo, Y., Earls, C.J., and Yulismana, W. (2004). Service load effective compression flange width in fiber reinforced polymer deck systems acting compositely with steel stringers. *Journal of Composites for Construction*, 8(4), 289–297.

Keller, T., and Gürtler, H. (2005). Composite action and adhesive bond between fiber-reinforced polymer bridge decks and main girders. *Journal of Composites for Construction*, 9(4), 360–368.

Machado, M.A.S., Sotelino, E.D., and Liu, J. (2008). Modeling technique for honeycomb FRP deck bridges via finite elements. *ASCE Journal of Structural Engineering*, 134(4), 572–580.Mainstone, R.J., and Menzies, J.B. (1967). Shear connectors in steel-concrete composite beams for bridges. 1. Static and fatigue tests on push-out specimens. *Concrete*, 1(9), 291–302.

Majumdar, P.K., Liu, Z., Lesko, J.J., and Cousins, T.E. (2009). Performance evaluation of FRP composite deck considering for local deformation effects. *Journal of Composite for Construction*, 13(4), 332–338.

Moon, F.L., Eckel, D.A., and Gillespie, J.W. (2002). Shear stud connections for the development of composite action between steel girders and fiber-reinforced polymer decks. *Journal of Structural Engineering*, 128(6), 762–770.

Oehlers, D.J., and Bradford, M.A. (1995). *Composite steel and concrete structural members*. Elsevier Science, New York.

Righman, J., Barth, K.E., and Davalos, J.F. (2004). Development of an efficient connector system for fiber reinforced polymer bridge decks to steel girders. *Journal of Composites for Construction*, 8(4), 279–288.

Slutter, R.G., and Fisher, J.W. (1966). *Fatigue strength of shear connectors*. Highway Research Record 147. Highway Research Board, Washington, DC, pp. 65–88.

5

FRP Deck–Steel Girder Bridge System

5.1 Overview

For conventional concrete deck-over-steel stringer bridges, full composite action is usually preferred and achieved due to the efficiency of the materials used. In the American Association of State Highway and Transportation Officials (AASHTO) code slab-on-girder bridges can be designed for either noncomposite or full composite actions. No partial composite action is defined. However, fiber-reinforced polymer (FRP) decks are usually designed in practice for partial composite action. Several limiting practical factors lead to this behavior: (1) The hollow core configuration of FRP panels and lack of continuous connection at the panel-stringer interface do not allow development of perfect contact and attachment between decks and connections. (2) The high modulus ratio between the steel girder and FRP panel (about 30 compared to 8–10 for a conventional concrete deck-over-steel girder) makes the contribution of the FRP deck to the overall bridge stiffness much less significant. (3) The practical connection spacing of about 0.6 to 1.2 m for FRP decks, compared to conventional concrete deck connection spacing of 0.15 to 0.25 m, is too large to develop full composite action. All these factors in turn lead to less shear force to be transferred at the deck-girder interface, leading to slippage and achieving a partial degree of composite action. On the other hand, it may actually be desirable to allow for some degree of deck-stringer relative displacement to account for differential thermal expansions between FRP and steel.

A number of design issues related to partial composite action in FRP deck systems need to be investigated, including: (1) transverse load distribution factors, (2) degree of composite action, (3) effective deck width, and (4) service limit and ultimate limit capacities, such as fatigue resistance and ultimate failure mode. Other design issues that are distinct for FRP decks include: (1) local deck deflections and (2) deck connection installation procedures. These issues will be studied in this chapter by experimental testing and verification by both finite element (FE) analysis and analytical method.

5.2 Experimental and FE Study on Scaled Bridge Model

5.2.1 Introduction

Although extensive research has been conducted on stiffness and strength evaluations of various types of FRP decks (Bakis et al. 2002; Davalos and Chen 2005; Chen and Davalos 2010), as shown in Chapters 2 and 3, only limited studies are available on FRP deck-on-steel girder bridge systems, which were mostly evaluated based on field- or lab-scale testing.

Keelor et al. (2004) conducted a field study on a short-span bridge located in Pennsylvania. This bridge had a pultruded FRP deck over five steel girders equally spaced at 1.75 m; the span length was 12.6 m and the deck thickness was 195 mm. The FRP deck was assumed to achieve full composite action through grouted stud connections welded to the stringers. Their results showed that under service load conditions, full composite action resulted in effective widths corresponding to about 90% for interior stringer spacing and 75% for exterior half stringer spacing.

Keller and Gürtler (2005) conducted lab tests on two large-scale T-sections to study composite action and effective flange width. Each test model was 7.5 m long with a pultruded FRP deck section of 1.5 m wide adhesively bonded to the top flange of a steel supporting beam. The in-plane normal strain distribution, i.e., strain parallel to the bonded surface, across the width of the FRP section was recorded at both upper and lower FRP facesheet components. The results showed that under the service limit state, the normal stress was almost uniform across the panel section. While under the failure limit state, the normal stress decreased toward the panel edges, indicating a more pronounced effect of shear lag under ultimate load. Later, two reduced scale T-sections were tested to service limit state and ultimate limit state (Keller and Gürtler 2005). One of the T-sections was fatigue loaded to 10 million cycles. The FRP pultruded flanges, which were adhesively bonded to the steel stringers, were able to achieve full composite action. At ultimate limit state, the failure mode was deck compression failure with the yielding of steel stringer. The structural behaviors of this full composite model were established at service and ultimate limit states. The deflection and ultimate strength could respectively increase by 30% and 56% by considering composite action. There were strain differentials between top and bottom facesheets due to low in-plane shear stiffness of the core. The strain distribution of top and bottom facesheets showed that the top facesheet fully participated as a top cord, while the bottom facesheet showed a more shear lag phenomenon. The effective flange width was smaller than that evaluated for a comparable concrete deck. Also, the T-section could sustain 10 million cycles of fatigue loading, which was comparable to the Eurocode 1 fatigue load on a reference bridge.

Fatigue tests on two T-beams with 7.5 m span length and adhesively bonded FRP deck to steel stringers were conducted by Keller and Tirelli

(2004). The pultruded FRP flange was fully and compositely connected to the steel stringers, and the FRP flange participated as the top cord of this T-section. The fatigue limit was considered to be 25% of static failure load at 10 million load cycles, which was far above the actual fatigue load in FRP bridges. The adhesive bond connection was proved to be able to sustain the fatigue loading.

The aforementioned studies were focused on FRP deck-on-steel girder systems with full composite action. However, in actual bridges, because of the low equivalent modulus of the FRP deck, the contribution of the deck to the bridge system is not as significant as for concrete, even for full composite action. Also, because of the difference in coefficient of expansion between the FRP and steel, where FRP has a typical coefficient of thermal expansion (CTE) in the range of 1.6 to $2.7 \times 10^{-5}/°C$ and steel has a CTE of $1.2 \times 10^{-5}/°C$, full composite action may induce an adverse effect on the bridge system if the shear connection is fully constrained. The prototype shear stud sleeve-type connection, as shown in Chapter 4, can address the above concerns. Since the practical spacing for this connection is about 0.6 to 1.2 m, compared to conventional concrete deck connection spacing of 0.15 to 0.25 m, only a partial degree of composite action can be achieved between the deck and stringers.

In the AASHTO standard (1996) and LRFD (2004) specifications, slab-on-girder bridges can be designed with either noncomposite or full composite action, but there is no provision for a partial degree of composite action. Bridges with partial composite action are typically designed as noncomposite, which is not economical since the contribution of the deck based on partial composite action is neglected. Therefore, this section will study the behavior of an FRP deck on a steel girder bridge system with a partial degree of composite action and provide design parameters, using a combined experimental investigation and FE analysis. In particular, the objectives are to (1) evaluate the performance of the prototype shear connection at the bridge system level, (2) evaluate the performance of the bridge system under static and fatigue loads, and (3) study the degree of composite action of the bridge system and its influence on the design parameters, including the transverse load distribution factor and effective flange width.

5.2.2 Test Plan

The test consisted of three phases. In phase I, static load tests were conducted on a 1:3 scaled bridge model to investigate transverse load distribution factors and local deflections of the FRP deck. In phase II, the same bridge model was subjected to a cyclic loading to evaluate the fatigue resistance of both shear connection and the bridge system. In phase III, a T-section was cut out from the bridge model and then loaded under three-point bending for linear and ultimate responses to investigate its effective flange width, degree of composite action, strength, and failure mode.

5.2.3 Test Models

5.2.3.1 Bridge Model Description

A 1:3 scaled bridge model with a span of 5.5 m was constructed consisting of three steel stringers (W16x36, Gr50) spaced at 1.22 m apart (Figures 5.1 and 5.2), based on a reference bridge designed according to AASHTO specifications (Zou 2008). A $5.5 \times 2.74 \times 0.13$ m FRP deck, consisting of three 1.8 m wide and 2.74 m long individual FRP honeycomb panels, as shown in Figure 1.1 and assembled by tongue-and-groove connections along the two 2.74 m transverse joints, was attached to the stringers using a prototype stud-sleeve connector. The longitudinal direction of the honeycomb core (Figure 1.1) is perpendicular to the traffic direction.

As shown in Figure 5.3 the FRP honeycomb panels consisted of top and bottom facesheets and a sinusoidal core. The facesheet had three layers of CDM 3208 laminate and two layers of chopped strand mat (ChSM) composed of E-glass fiber and polyester resin, with the material properties shown in Tables 5.1 and 5.2. Another layer of ChSM (0.256 kg/m^2) was placed in between the facesheet and core as a bonding layer. The total thickness of the FRP panel was 125 mm, where the facesheet and the core were 12.5 and 100 mm thick, respectively.

FIGURE 5.1
Photo of scaled bridge model.

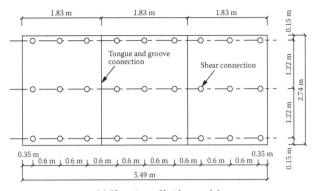

(a) Plan view of bridge model

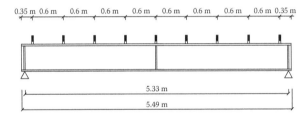

(b) Elevation view of bridge model

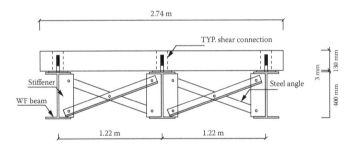

(c) Cross section of bridge model

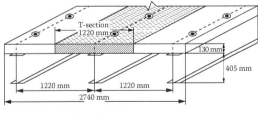

(d) 3-D bridge model

FIGURE 5.2
Details of scaled bridge model.

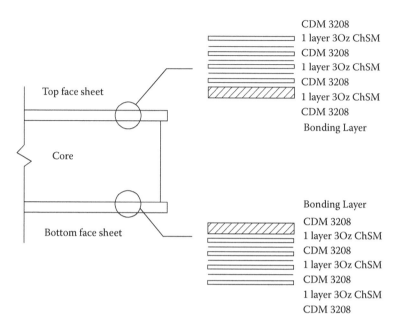

FIGURE 5.3
Facesheet lay-up.

TABLE 5.1

Properties of Constituent Materials

Material	E, GPa (×10⁶ psi)	G, GPa (×10⁶ psi)	v	ρ, g/cm³ (lb/in.³)
E-glass fiber	72.4 (10.5)	28.8 (4.18)	0.255	2.55 (0.092)
Polyester resin	5.06 (0.734)	1.63 (0.237)	0.30	1.14 (0.041)

TABLE 5.2

Material Properties of Facesheet

		Nominal Weight (kg/m²)	Thickness (mm)	V_f
CDM 3208	0°	0.531	0.49	0.424
	90°	0.601	0.55	0.425
	ChSM	0.256	0.25	0.396
Bonding layer	ChSM	0.915	1.91	0.188

During construction, the steel girders with prewelded shear studs at 0.6 m apart were first erected and braced at the supports and mid-spans. FRP panels, with 80 mm circular holes cut at corresponding locations, were installed on the top of the steel girders, and then snug-fitted by transverse tongue-and-groove connections, which were joined using polymer resin and bonded at the top and along the joints using fiber glass sheets, as shown in Figure 5.4. The tongue-and-groove honeycomb core connection was about 0.15 m wide and was filled with polymer concrete to strengthen the joint, as shown in Figure 5.5. Finally, the shear connections were installed, as shown in Figure 5.6.

5.2.3.2 T-Section Model Description

After completing the tests on the scaled bridge model, a T-section was cut out with a flange width of 1.22 m, as illustrated in Figures 5.2 and 5.7, which was supported by a steel girder. Three brackets were placed on each side of the flange to provide lateral support to the flange section.

FIGURE 5.4
Tongue-and-groove connection with FRP sheet covered.

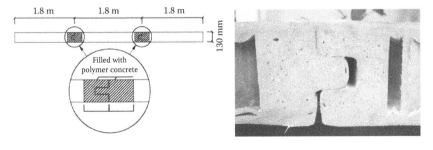

FIGURE 5.5
Tongue-and-groove connections.

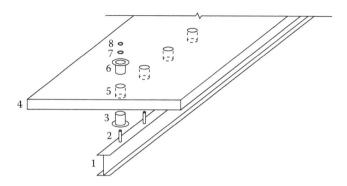

FIGURE 5.6
Installations of shear connection to FRP decks.

FIGURE 5.7
T-beam test model.

5.2.4 Test Procedures

5.2.4.1 Phase I Test—Static Behavior of Bridge Model

Three load cases were used for the bridge model (Figure 5.1) with shear connection spacing of either 0.6 or 1.2 m, as shown in Figure 5.8 and Table 5.3. The model was first tested with shear connections at 0.6 m intervals. The 1.2 m spacing was achieved by removing the top sleeves of every other shear connections. Load case 1 (LC1, Figure 5.8) was designed to study the transverse load distribution factor. A concentrated load was applied over an area of 0.6 × 0.25 m, using a 245 kN actuator to simulate truck wheel load, at

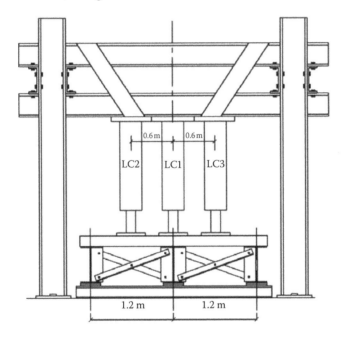

FIGURE 5.8
Load cases on scaled bridge model.

TABLE 5.3

Load Case Designation

Load Case	Transverse Load Position	Connection Spacing (m)	Label
1	Aligned with girder 2	0.6	LC1
1	Aligned with girder 2	1.2	LC1
2	Aligned with mid-point of girders 1, 2	0.6	LC2
2	Aligned with mid-point of girders 1, 2	1.2	LC2
3	Aligned with mid-point of girders 2, 3	0.6	LC3
3	Aligned with mid-point of girders 2, 3	1.2	LC3

mid-span and over the middle girder, as shown in Figure 5.8. The bridge was loaded to 50% service limit load by a displacement control rate of 1 mm/min. The 50% service load was based on the level calculated for the reference full-scale bridge. Based on AASHTO specification (2004), the service stress limit for a service II load combination can be calculated as

$$f_f = 0.8 R_h F_{yf} \tag{5.1}$$

where R_h is the hybrid factor, which is 1.0 for the homogenous section used in this model, and F_{yf} is the yield strength of the flange. Therefore, the service stress limit f_f was 275 MPa (0.8 F_{yf}) for grade 50 (F_{yf} = 344 MPa) steel girders used in this model, and the model was loaded until the flange stress reached 138 MPa (0.5 f_f), corresponding to a 178 kN patch load.

Load cases 2 and 3 were used to study the local deflection of the FRP deck (Figure 5.8). A 36 kN patch load, corresponding to a rear wheel load of an HS20 truck, was respectively applied at the mid-point between girders 1 and 2 and girders 2 and 3 (Figure 5.8). Since load cases 2 and 3 were symmetrical, the local deflection of the deck was calculated as the average values of these two load cases.

A transverse deflection profile was obtained by the measurements from five LVDTs across the mid-span section, as shown in Figure 5.9, where LVDTs 1 through 3 were placed directly under girders 1 through 3, respectively, and LVDTs 4 and 5 were placed under the FRP deck at mid-span. The local deflection was defined as the relative displacement of the panel between the two supporting girders by linear interpolation.

5.2.4.2 Phase II Test—Fatigue Behavior of Bridge Model

Similar to the static test, a patch load was applied at the mid-span of the middle girder. The load range was 0–112 kN. The model was subjected to 10.5

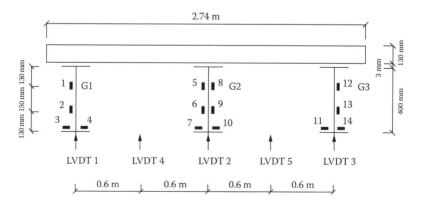

FIGURE 5.9
Instrumentation of bridge model.

million cycles loading, which was equivalent to 75 years bridge service life span (Moon and Gillespie 2005). The fatigue load was calculated as comparable to the corresponding design load for a reference full-scale bridge, where the stress range was 3.3 MPa for induced shear stress at each shear connection (Zou 2008), which corresponded to a cyclic load of 0–25.1 kips at the mid-span of the center girder. The loading was stopped at every 2 million cycles, and the static deflection of the model was measured to calculate the stiffness.

5.2.4.3 Phase III Test—T-Beam Behavior

A patch load was applied at the mid-span of the T-beam over an area of 0.6×0.25 m² using a 490 kN actuator to study its effective flange width, as shown in Figure 5.7. The system was subjected to three-point bending loading with displacement control at a rate of 1 mm/min within the service load limit, and the load-displacement relation was recorded. A total of 20 strain gages were attached at the top and bottom surfaces of the deck, with 10 at quarter span and 10 at mid-span, as shown in Figure 5.10, to measure the longitudinal normal strain of the FRP flange and evaluate the effective flange width.

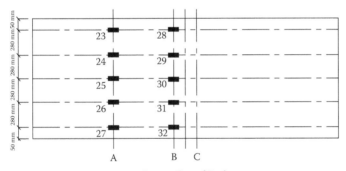

Bottom Face of Deck

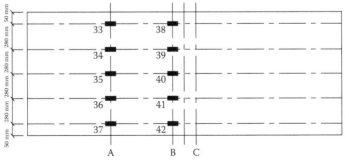

Top Face of Deck

FIGURE 5.10
Instrumentation of T-beam FRP deck section.

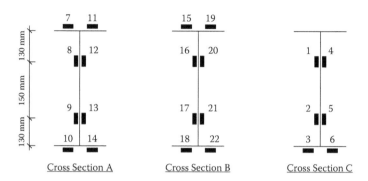

FIGURE 5.11
Instrumentation of T-beam girder (see Figure 5.10 for locations of cross sections).

In addition, strain gages were bonded along the depth of the girder to measure the longitudinal normal strains, as shown in Figure 5.11. Based on the strain distributions, the neutral axis of the system can be determined, and subsequently, the degree of composite action can be calculated, as will be shown in the next section.

5.2.5 Test Results

5.2.5.1 Phase I Test Results of Bridge Model

5.2.5.1.1 Load Distribution Factor

As suggested by Eom and Nowak (2001), the expression

$$LDF = \frac{M_i}{\sum\limits_{j=1}^{N} M_j} = \frac{ES_i \varepsilon_i}{\sum\limits_{j=1}^{N} ES_j \varepsilon_j} = \frac{\dfrac{S_j}{S_l}\varepsilon_i}{\sum\limits_{j=1}^{N} \dfrac{S_j}{S_l}\varepsilon_j} = \frac{\varepsilon_i w_i}{\sum\limits_{j=1}^{N} \varepsilon_j w_j} \tag{5.2}$$

is adopted to obtain load distribution factor (LDF), where M_i = bending moment at the ith girder, E = modulus of elasticity, S_i = section modulus of the ith girder, S_l = typical interior section modulus, ε_i = maximum bottom-flange static strain of the ith girder, which can be obtained from the test results, w_i = ratio of the section modulus of the ith girder to that of a typical interior girder, and N = number of girders. If the girders have the same section modulus, which is the case in this study, (5.2) can be simplified as

$$LDF = \frac{\varepsilon_i}{\sum\limits_{j=1}^{N} \varepsilon_j} k \tag{5.3}$$

where k is the number of lanes loaded.

TABLE 5.4

Load Distribution Factor (LDF) of Test Model

	LDF from AASHTO Specifications	Test Results			
		0.6 m Connection Spacing		1.2 m Connection Spacing	
		LDF	Difference (%)	LDF	Difference (%)
		0.647	—	0.644	—
AASHTO standard	0.727	—	12.3%	—	12.9%
AASHTO LRFD	0.655	—	1.2%	—	1.7%

Based on (5.3), the load distribution factors under the service load for the middle girder were calculated to be 0.647 and 0.644, respectively, for the 0.6 and 1.2 m connection spacing, as shown in Table 5.4. The values calculated from AASHTO standard specification (1996) and AASHTO LRFD specifications are also listed in Table 5.4. It can be seen from Table 5.4 that the load distribution factor from the test result with a connection spacing of 0.6 m is 1.2 and 12.3% higher than values from AASHTO LRFD and AASHTO standard specifications, respectively. Although the two specifications do not consider bridges with a partial degree of composite action, it seems that they can still provide reasonable accuracy in predicting the load distribution factor for a bridge system that has a partial degree of composite action.

5.2.5.1.2 Local Deck Deflection

As shown in Figure 5.12, the induced deflection profile clearly displays the localized effect, where the largest deflection occurs at the loading position. The local deflections were calculated to be 1.65 mm (L/727) and 1.75 mm (L/686) for the connection spacing of 0.6 and 1.2 m, respectively, as shown

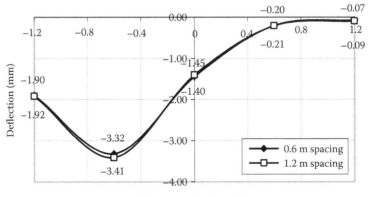

FIGURE 5.12
Deflection profile corresponding to load case 2 (see Figure 5.8 for load case definition).

TABLE 5.5

Deflection Profile of Test Model

LVDT # (see Figure 5.8)	Deflection (mm)	Local Deflection (mm)
0.6 m Connection Spacing		
1	1.91	
4	3.31	1.65
2	1.45	
5	0.20	
3	0.076	
1.2 m Connection Spacing		
1	1.93	
4	3.40	1.75
2	1.40	
5	0.20	
3	0.10	

in Table 5.5. In AASHTO LRFD specifications (2004), there is no deflection limit for FRP bridge decks, although there is a provision for an orthotropic bridge deck, which is intended for steel orthotropic deck with ribs. However, some researchers suggested L/400 as the deflection limit (Demitz et al. 2003; Zhang and Cai 2007). Therefore, the local deflection for this FRP panel is considered to be acceptable.

5.2.5.2 Phase II Test Results of Bridge Model

5.2.5.2.1 Fatigue Resistance

No apparent damage on the shear connection and the bridge system was observed for the static test. For the fatigue test, the stiffness remained nearly constant throughout the loading history, as shown in Figure 5.13, which was not surprising as the applied stress range was much lower than the threshold value obtained from the single-connector push-out test of 53 MPa (Davalos et al. 2010). The test results indicated that the shear connection and the bridge system can meet both strength and fatigue requirements established by the AASHTO LRFD specifications (2004).

5.2.5.3 Phase III Test Results of T-Beam

5.2.5.3.1 Degree of Composite Action

Based on the strain distributions from the test, with one example shown in Figure 5.14 at 100% service load, the neutral axis of the system, which determines the degree of composite action, can be plotted as in Figure 5.15.

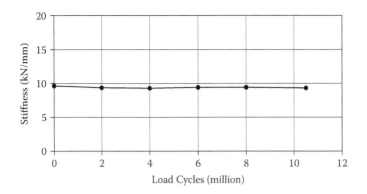

FIGURE 5.13
Stiffness ratio variations during fatigue test.

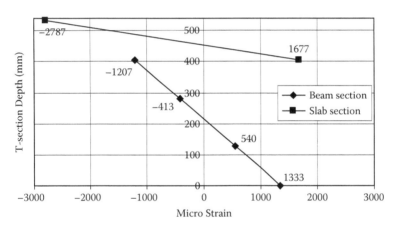

FIGURE 5.14
Axial strain distribution at mid-span cross section B (see Figure 5.10 for locations of cross sections).

Subsequently, the degree of composite action (DCA) can be calculated as (Park et al. 2005)

$$DCA(\%) = \frac{N_p - N_0}{N_{100} - N_0} 100 \qquad (5.4)$$

where N_0 and N_{100} are predicted neutral axes corresponding respectively to non- and full composite actions, which can be calculated respectively as 202 and 256 mm (Zou 2008), and N_p is the neutral axis for partial composite action, which is about 213 to 218 mm from Figure 5.15. Thus, the degree of composite action of the system can be calculated to be about 25%. It is interesting to observe that the effect of connection spacing is only marginal for

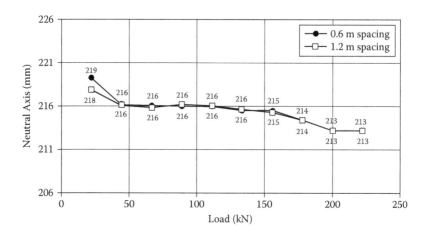

FIGURE 5.15
Neutral axis position.

both 0.6 and 1.2 m spacing, as the neutral axes are very close, as shown in Figure 5.15. Further study is necessary to evaluate the effect of spacing on the degree of composite action.

5.2.5.3.2 Effective Flange Width

The strain profiles at mid-span are plotted for both top and bottom surfaces in Figures 5.16 and 5.17, respectively. These strain values are curve-fitted by approximate functions. In bridge engineering, the three-dimensional behavior of a bridge system is usually reduced to the analysis of a T-beam section with a reduced width of deck in relation to center-to-center spacing of stringers, over which the longitudinal normal stresses are assumed to be uniformly distributed, which is termed effective flange width. This concept will be further discussed in Section 5.3. Based on definition of effective flange width, the longitudinal normal stress is assumed to be uniformly distributed along the panel section. The effective flange width can be expressed as the integral of normal stress distribution divided by the maximum stress at the panel-stiffener intersection as

$$b_e = \frac{\displaystyle\int_{-b/2}^{b/2} \sigma_x \, dx}{\sigma_{max}} = \frac{2\displaystyle\int_{0}^{b/2} \sigma_x \, dx}{\sigma_{max}} \tag{5.5}$$

The average normal stress, σ_x, which can be assumed uniform along its segmentally discrete width h, can be obtained from the results of the strain gages. The integral of the in-plane normal stresses, i.e, normal stress parallel

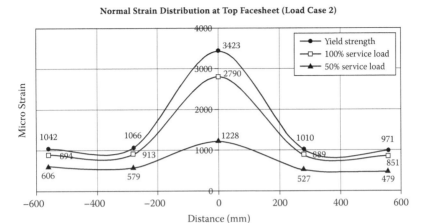

FIGURE 5.16
Normal strain distribution on top facesheet of FRP panel.

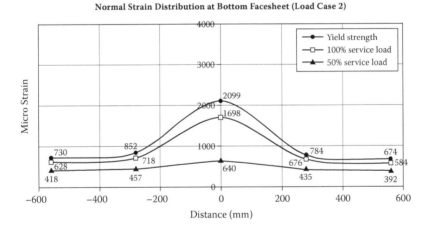

FIGURE 5.17
Normal strain distribution on bottom facesheet of FRP panel.

to the deck, can be approximated as the summation of the discrete values over the panel section, which can be expressed as

$$\int_{-b/2}^{b/2} \sigma_x \, dx = \sum_{i=1}^{n} \sigma_{x,i} h_i \qquad (5.6)$$

where $\sigma_{x,i}$ is the normal stress calculated from the normal strain at strain gage i, σ_{max} is the stress value at the panel-girder centerline, h_i is the width for each block, and n is the total number of blocks.

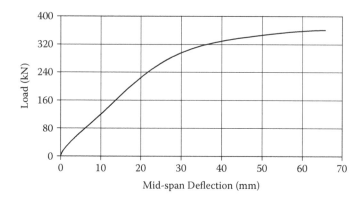

FIGURE 5.18
Load-deflection curve.

Based on (5.5) and (5.6), the effective flange width was calculated to be about 0.63 m, which is about 50% of the actual flange width. Therefore, a reduction factor needs to be introduced to calculate the effective flange width for bridges with a partial DCA vs. the effective flange width for full composite action, which will be discussed in Section 5.3.

5.2.5.3.3 Failure Mode at Strength Limit

The T-section displayed nearly linear elastic behavior until the load reached about 220 kN, and then the load-deflection curve showed a shallow nonlinear behavior, as shown in Figure 5.18. The maximum load was about 356 kN. The T-section failed by local buckling of the flange of the steel girder, as shown in Figure 5.19, which indicated that the FRP deck cannot brace the steel girder as effectively as a concrete deck. This also corroborates the finding of partial composite action (estimated as 25%), as described before.

5.2.6 Finite Element Model

An FE model was created to simulate the tests described above using ABAQUS (Dassault Systèmes 2007), as shown in Figure 5.20. The model consisted of three parts: steel stringers, FRP deck, and shear connections.

The steel was modeled as a linear elastic isotropic material with the modulus of elasticity equal to 200,000 MPa and the Poisson's ratio equal to 0.3. The elements used consisted of four-node general purpose shell elements, with reduced integration, hourglass control, and finite membrane strains. These elements in ABAQUS are commonly referred to as S4R (Dassalt Systems 2007). The cross-bracings were modeled by two-node beam elements.

As shown in Figure 5.3, the FRP deck consisted of top and bottom facesheets, and a core with sinusoidal shape. It is demanding to model the actual geometry of the FRP deck. Therefore, the FRP deck was simplified into an equivalent FRP plate. The properties of the FRP plate, as shown in

FIGURE 5.19
T-section prior to failure.

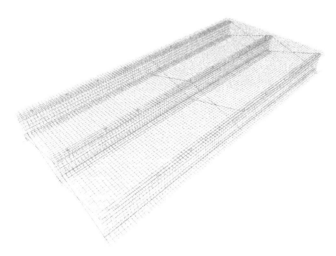

FIGURE 5.20
FE bridge model.

Table 5.6, were obtained using a stiffness analysis based on homogenization theory as shown in Chapter 2 and verified by bending tests on discrete FRP panels, as shown in Section 5.2.9. The same S4R elements were used to model the equivalent FRP plate.

The most important characteristic of this model is to simulate the interaction between the FRP deck and the steel girder. This was achieved using multiple-point constraint connector elements (CONN3D2 in ABAQUS) among the

TABLE 5.6

Equivalent Properties of FRP Panel

	E_x (Mpa)	E_y (Mpa)	v_x	G_{xy} (Mpa)
In-plane	2,560	2,300	0.303	560
Bending	5,640	5,640	0.303	1,400

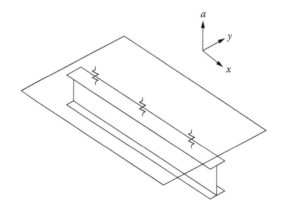

FIGURE 5.21
Illustration of connector element.

common nodes over the center contact section between the top flange of the beam and the bottom of the deck, as shown in Figure 5.21. Each connector element extended between the centerline of the beam flange thickness and the centerline of the deck thickness. The following boundary conditions for the connector element were specified: (1) all three rotational degrees of freedom are coupled between the beam and the deck (no relative rotation), (2) the vertical displacement in the z direction is coupled (no relative displacements), and (3) elastic displacements are prescribed in the x and y directions, by varying the elastic displacement constants, to simulate partial composite behavior between the deck and the girder, where the stiffness was 1.46 kN/mm based on the testing results from Chapter 4 and Davalos et al. (2011).

The patch load on the physical test specimen consisted of a 600 × 250 mm area, which was applied as a pressure load over the elements. Pin-roller constraint was used to represent the simply supported boundary conditions from the test.

5.2.7 FE Analysis Results

Figure 5.22 displays the displacement contour of the FRP panel subjected to load case 2 as shown in Figure 5.8. The deflections are shown in Table 5.7, where good correlations can be observed between the FE and testing results as shown in Table 5.5. Using the data reduction techniques as shown in (5.3)

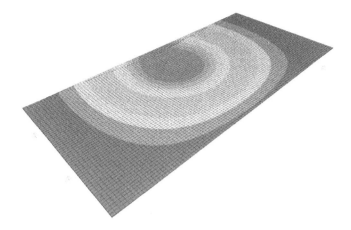

FIGURE 5.22
Deformed shape of panel.

TABLE 5.7

Deflection Profile of Bridge Model

Deflection Point	Deflection (mm)	Local Deflection (mm)
0.6 m Connection Spacing		
1	1.77	
2	2.81	1.12
3	1.54	
4	0.45	
5	0.12	

and (5.5), the LDF and effective flange width for the connection spacing at 0.6 m are 0.621 and 0.75 m, respectively, which are close to the testing results of 0.647 and 0.63 m, illustrating the relative accuracy of the FE model formulated herein. The FE model can be used to carry out parametric studies.

5.2.8 Conclusions

In this section, a one-third scaled bridge model consisting of an FRP sandwich deck attached to three steel girders by a mechanical connector was constructed and tested. A T-beam section was cut out from the scaled bridge to be tested under three-point bending until failure. An FE model was constructed to simulate the tests. Since the practical spacing for the connectors was about 0.6 to 1.2 m, compared to conventional concrete deck connection spacing of 0.15 to 0.25 m, there is a partial degree of composite action between the deck and stringers. In the AASHTO standard (1996) and LRFD (2004) specifications, slab-on-girder bridges can be designed with either noncomposite or full composite action, but there is no provision for a partial degree of composite action. Therefore, this section was directed to study the effect of the partial composite action on the performance of the bridge system. Based on this study, the following conclusions can be drawn:

1. The bridge system and shear connections were able to sustain a cyclic fatigue loading equivalent to a 75-year bridge service life span, and showed nearly no stiffness or strength degradation for both static and fatigue tests. Therefore, the prototype shear connection used can adequately perform at a bridge system level and be applicable in practice. About 25% degree of composite action can be achieved with the prototype shear connection.

2. It is possible to use AASHTO LRFD specifications (2004), and AASHTO standard specifications (1996) can be used to predict load distribution factors for FRP decks on steel girder bridges with partial degree of composite action, where the former provides better predictions for the bridge system considered in this study.

3. The effective flange width for bridges with partial DCA can be calculated from the effective flange width for full composite action, with the introduction of a reduction factor *R*, for a specified percent of composite action, which will be discussed in Section 5.3.

4. The two different connection spacings used in the present model, 0.6 and 1.2 m, do not have a significant impact on the structural behavior or performance, such as load distribution factor or degree of composite action, for the FRP deck on the steel girder system considered in this study. Further study is required to evaluate the effect of connection spacing.

5. The panel local deflection ratio is about L/700, which is considered to be acceptable. The failure mode of the bridge system was local buckling of the top flange of the steel girder, while the deck remained relatively intact.

6. Good correlations can be observed between the testing and FE results.

5.2.9 Evaluation of FRP Panel Properties

5.2.9.1 Equivalent Material Properties Based on Homogenization Theory

Equivalent properties of the FRP panel were obtained following the method presented in Chapter 2. First, the layer properties of the facesheet were calculated based on constitute materials as shown in Table 5.1 and the volume fractions from Table 5.2, with the results shown in Table 5.8. The stiffness of the facesheet was then predicted using lamination theory, as shown in Table 5.9. Stiffness evaluation of the core was based on a homogenization concept and mechanics of materials approach, with the results shown in Table 5.9. Finally,

TABLE 5.8

Stiffness Properties of Facesheet Lamina

	E_1 (MPa)	E_2 (MPa)	G_{12} (MPa)	G_{23} (MPa)	v_{12}	v_{23}
CDM 3208	35,900	11,100	2,810	3,030	0.305	0.509
CDM 3208 CSM	17,400	17,400	6,200	6,200	0.406	0.406
Bonding layer	9,820	9,820	3,510	3,510	0.397	0.397

TABLE 5.9

Stiffness Properties of Facesheet and Core

	E_x (MPa)	E_y (MPa)	v_x	G_{xy} (MPa)
Facesheet	13,600	14,100	0.304	3,500
Core	530	1.0	0.431	0.7

the equivalent 2D orthotropic properties of the sandwich panel were calculated by modeling it as a three-layer laminated system of top and bottom faces and core using the lamination theory, with the results shown in Table 5.10.

5.2.9.2 Test Setup

To evaluate the accuracy of the equivalent properties obtained in Table 5.10, the three FRP panels, which were used as a deck for the scaled bridge, were tested to obtain their stiffness prior to the installation of the scaled bridge. As described earlier, the panels were 1.8 m wide (transverse direction as shown in Figure 1.1) and 2.74 m long (longitudinal direction as shown in Figure 1.1). As shown in Figure 5.23, the panels were supported along the longitudinal direction. A patch load was applied using a 245 kN actuator over an area of 0.6×0.25 m^2, as shown in Figure 5.23. The panels were loaded up to 45 kN with force control at a loading rate of 9 kN/min at five different locations, numbered 1 through 5, as shown in Figure 5.24. For panel 1, 10 unidirectional strain gages were bonded at the five locations to measure longitudinal and transverse strains. Five LVDTs were installed below the deck to measure vertical displacements at these locations. For panels 2 and 3, only the displacements at the center of the panel, i.e., at location 3, were measured.

TABLE 5.10

Equivalent Properties of FRP Panel

	E_x (MPa)	E_y (MPa)	v_x	G_{xy} (MPa)
In-plane	2,940	2,641	0.303	648
Bending	6,488	6,488	0.303	1,600

FIGURE 5.23
Panel test setup.

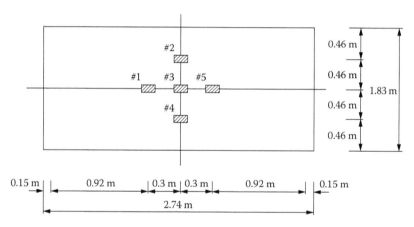

FIGURE 5.24
Loading positions and instrumentations.

5.2.9.3 Test Results

The measured deflections and strains are presented in Tables 5.11 and 5.12. The units for the displacement and strain are mm/kN and microstrain/kN, respectively. L represents longitudinal strain, and T represents transverse strain. The transverse strain gage at location 4 failed and no strain data were available.

5.2.9.4 FE Analysis

The testing results as shown in Tables 5.12 and 5.13 cannot be directly used to evaluate the accuracy of the equivalent properties as shown in Table 5.10.

TABLE 5.11

Test Results—Deflections

Test	Deflection (10^{-2}*mm/kN)				
Load Case	1 (LVDT1)	2 (LVDT2)	3 (LVDT3)	4 (LVDT4)	5 (LVDT5)
1	−15.3	−15.4	−15.9	−15.7	−12.6
2	−14.2	−24.4	−17.5	−12.8	−15.4
3	−16.7	−18.2	−19.8	−18.3	−16.9
4	−14.1	−12.4	−17.5	−25.0	−15.3
5	−12.2	−15.7	−16.1	−16.1	−16.3
Panel 2	−15.0	−18.8	−20.6	−19.4	−17.4
Panel 3	−17.1	−19.6	−21.5	−20.0	−18.4

TABLE 5.12

Test Results—Longitudinal and Transverse Strains

Test	Longitudinal Strain (microstrain/kN)				
Load Case	1L	2L	3L	4L	5L
1	19.5	13.8	9.6	13.2	8.4
2	15.7	27.1	16.6	12.5	15.2
3	13.2	19.4	24.1	19.9	12.9
4	12.4	13.1	14.8	25.4	11.3
5	8.9	14.6	9.1	12.8	18.9
Test	Transverse Strain (microstrain/kN)				
Load Case	1T	2T	3T	4T	5T
1	13.9	2.0	7.0	N/A	2.5
2	2.4	10.4	3.1	N/A	2.1
3	6.2	2.8	14.8	N/A	6.2
4	1.7	3.8	2.5	N/A	1.7
5	1.4	1.2	6.8	N/A	13.5

TABLE 5.13

Deflection Correlations between Test and FE Results

Location	LC 1		LC 2		LC 3		Panel 2		Panel 3	
	Test	FE	Test	FE	Test	FE	Test	FE	Test	FE
1	15.9	14.0	14.2	13.6	16.7	13.8	15.0	13.8	11.4	13.8
2	15.5	13.6	24.7	23.4	18.2	16.8	18.8	16.8	19.6	16.8
3	16.0	13.8	17.5	16.8	19.8	18.2	20.6	18.2	21.5	18.2
4	15.9	13.6	12.6	11.9	18.3	16.8	19.4	16.8	20.0	16.8
5	12.4	10.2	15.4	13.6	16.9	13.8	17.4	13.8	18.4	13.8
Difference (%)	16.7		6.7		13.9		15.0		13.9	

Therefore, an FE model was constructed to simulate the bending test. Similar to the FE model for the scaled bridge, the FRP panel was modeled as an equivalent FRP plate using S4R shell elements, with 2D orthotropic properties as shown in Table 5.10. The pin-roller boundary condition and loading conditions as those used in the test were simulated in the FE model. Since load cases 1 and 2 are symmetric to load cases 4 and 5, only results for load case 1 through 3 are reported in Tables 5.13 and 5.14 for deflections and strains, respectively.

5.2.9.5 Correlations between Testing and FE Results

As can be seen from Tables 5.11 and 5.12, although load cases 1 and 2 are symmetric to load cases 4 and 5, the deflections and strains obtained from the tests are not symmetric with minor differences, due to geometric imperfections and manufacturing errors of the samples. Therefore, average values of load cases 1 and 5, and load cases 2 and 4 are shown in Tables 5.13 and 5.14.

As can be seen from Tables 5.13 and 5.14, the differences between the testing and FE results are consistent for both deflections and strains. For deflection, the differences between FE and testing range from 6.7 to 16.7%, with an average value of 13%. For strain, the difference between test and FE analysis varies from 7 to 20%, with an average value of 13%, except for load case 2, since the strains are small in the transverse direction. To compensate for the differences between the testing and FE model, the equivalent properties, as shown in Table 5.12, are reduced to 87%, as shown in Table 5.13, which are subsequently used in the FE model of the scaled bridge.

TABLE 5.14

Strain Correlations between Test and FE Results

Strain Location	LC 1		LC 2		LC 3	
	Test	FE	Test	FE	Test	FE
1L	19.2	18.3	14.0	11.4	13.2	12.1
2L	14.2	11.7	26.3	24.8	19.4	16.3
3L	9.4	12.1	15.7	16.1	24.1	20.8
4L	13.0	11.7	12.8	10.6	19.9	16.3
5L	8.6	7.1	13.2	11.4	12.9	12.1
Difference (%)	7.7		12.6		14.7	
1T	13.7	12.7	2.1	1.2	6.2	5.0
2T	1.6	2.2	10.4	10.1	2.8	2.3
3T	6.9	5.0	5.3	0.9	14.8	13.0
4T	N/A	2.2	−3.8	−0.7	N/A	2.3
5T	1.9	1.7	1.9	1.2	6.2	5.0
Difference (%)	7.2		207.3		20.7	

5.3 Evaluation of Effective Flange Width by Shear Lag Model

5.3.1 Introduction

In bridge engineering, a deck-and-girder system acting compositely is usually reduced to the analysis of a T-beam section characterized by the elementary beam theory, and called beam-line analysis. In this approach, and consistent with beam theory, the longitudinal normal stress on the deck section is assumed to be constant. However, due to in-plane shear flexibility of the deck, the longitudinal normal stress over a center-to-center bridge deck section is nonuniform along its transverse cross section, with the maximum value occurring at the mid-line junction with the girder, and gradually decreasing toward the center spacing line. This nonuniform distribution of stress is known as shear lag. The stress distribution depends on several factors, such as cross-sectional dimensions and stiffness of the deck (flange) and girder and loading conditions, resulting in analytical solutions not easily applicable in practice. Therefore, an effective flange width is used in design practice to simplify the problem. The effective flange width is defined as a reduced width of deck, in relation to center-to-center spacing of girders, over which the normal or longitudinal stresses are assumed to be uniformly distributed, based on the premise that the stress resultant over the effective width should be equal to the stress resultant over the actual flange width. This concept was adopted in both AASHTO standard (1996) and AASHTO LRFD specifications (2004), defined primarily for concrete decks in composite steel bridges. The Canadian highway bridge design code CSA (2000) defined the effective flange width in a similar manner as AASHTO (1996), with formulas primarily developed by Cheung and Chan (1978). Both AASHTO and Canadian specifications consider span length and girder spacing as the most important parameters that can affect effective flange width. The AASHTO LRFD code additionally includes slab thickness and girder dimensions.

Numerous studies have been conducted on this topic, most of which were focused on concrete decks. Moffat and Dowling (1975) studied the effective flange width for steel box girder bridges using FE analysis. The bridges were loaded with both uniform load and point load. They concluded that the ratio of girder spacing to span length was the most significant factor, and loading types and positions were other factors that affected the effective flange width. Moffat and Dowling (1978) later studied the effective flange width provisions in the British bridge code and pointed out that the nondimensional ratio of flange width to span length was the most dominant parameter, and girder size and deck thickness had little effect on the effective flange width for most practical bridges. Cheung and Chan (1978) used the finite strip method to study a wide range of steel bridges and box girder bridges. It was also concluded that girder spacing and bridge span length were major factors, while slab thickness and girder sections had little effect on effective flange width. Also, the effective flange width was found to be independent of the number of traffic lanes, and upper and lower bound values could be provided from models for multiple girder bridges under a uniformly distributed load and single T-beam sections under a point load, respectively.

Ahn et al. (2004) used a simply supported reference bridge to compare the values of effective flange width from several design specifications/codes, including AASHTO, BS5400, Canadian code, Japanese code, and Eurocode 4. Amadio et al. (2004) evaluated the effective flange width according to Eurocode 4 at ultimate strength state, by testing four composite T-beams until failure. They found that the effective flange width approached the whole slab width when the T-beam was close to failure. They concluded that it was conservative to define the same effective flange width for both service limit and ultimate limit states, but provided separate formulas for service limit and ultimate limit states.

For analytical solutions, Adekola (1968) developed a method that accounted for both plane stress and bending stress effects in the shear lag phenomenon, where the effective flange width was subsequently divided into the shear flange effective width and bending effective flange width. Adekola (1974) proposed a more rational basis for defining effective flange width, based on girder deflection rather than flange stress. In his definition, the deflection response of the equivalent T-section was the same as that of the actual T-section. By adopting this new definition, he studied the shear lag phenomenon with a partial interaction for concrete deck-on-steel stringer bridges, and the results showed that effective flange width increased with an increase in the degree of interaction or composite action. Song and Scordelis (1990a, 1990b) conducted harmonic shear lag analysis using plane stress for the flanges of simple or continuous beams with different girder cross sections, and he presented simplified empirical formulas and diagrams for determining the shear lag effects in simple beams under various loading conditions.

For timber bridges, Davalos and Salim (1993) studied effective flange width for 125 stress-laminated timber bridges by FE analysis. Empirical equations for effective flange width were proposed, considering major variables, including girder spacing, bridge span length, ratio of girder depth to deck thickness, and ratio of longitudinal girder elastic modulus to deck elastic modulus.

There are only limited studies available for effective flange width of FRP deck panels accounting for orthotropic behavior. Tenchev (1996) conducted an FE parametric study on the shear lag phenomenon for orthotropic plates, based on which empirical equations for effective flange width were proposed. Keelor et al. (2004) conducted a field study on a short-span bridge in Pennsylvania. The bridge was 12.6 m long with five steel girders equally spaced at 1.8 m, and had a pultruded FRP deck with a thickness of 19.5 cm, and the design was based on full composite action. Their results showed that under service load conditions, the effective flange widths corresponded to about 90% of girder spacing for interior girders, and 75% of half girder spacing for exterior girders, respectively. Keller and Gürtler (2005) conducted lab tests on two large-scale T-sections to study composite action and effective flange width. Each test model was 7.5 m long with a pultruded FRP deck section of 1.5 m wide adhesively bonded to the top flange of a steel supporting beam. The normal strain distribution across the width of the FRP section was recorded at both top and bottom FRP facesheets. The results showed that under service limit state, the normal stress was almost uniform across the panel section. While under failure limit state, the normal stress decreased toward the panel edges, indicating a more pronounced effect of shear lag.

Considering the increasing field implementation of FRP decks, it would be advantageous to develop an analytical model to calculate effective flange width for orthotropic bridge decks, which is the objective of this section. Using a harmonic analysis developed for FRP thin-walled sections (Salim and Davalos 2005), a shear lag model is proposed, as described next.

5.3.2 Shear Lag Model

For an orthotropic deck supported by steel girders with full composite action, the deck panel can be assumed to be stiffened by the two girders, as shown in Figure 5.25. The following assumptions are adopted in this model to simplify the problem: (1) the axial force N_y and moment M_y are assumed to be zero, and (2) the twisting moment in the plate is neglected ($M_{xy} = 0$). When the bridge is subjected to out-of-plane load, the stress distribution due to out-of-plane moment is shown in Figure 5.26a, where the deck is under compression and the steel girder is under tension, and the compression force in the deck is transferred through shear connections at the interface between the deck and steel girder. Therefore, only edge shear tractions N_{xy} and axial force N_x are acting on the panel, as shown in Figure 5.27, and the constitutive and compliance matrices can be given as

$$\left\{ \begin{array}{c} N_x \\ N_{xy} \\ M_x \end{array} \right\} = \left[\begin{array}{ccc} A_{11} & 0 & 0 \\ 0 & A_{66} & 0 \\ 0 & 0 & D_{11} \end{array} \right] \left\{ \begin{array}{c} \varepsilon_x \\ \gamma_{xy} \\ \kappa_x \end{array} \right\} \tag{5.7}$$

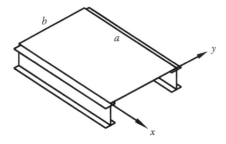

FIGURE 5.25
Typical panel element with two stiffened edges.

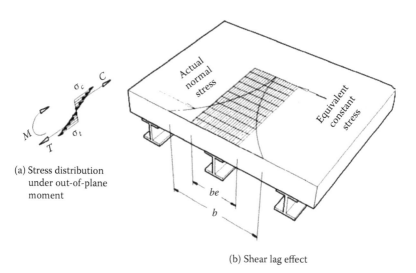

(a) Stress distribution
    under out-of-plane
    moment

(b) Shear lag effect

FIGURE 5.26
Effective flange width.

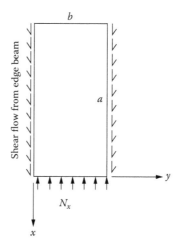

FIGURE 5.27
Loading condition for flange element.

$$\left\{ \begin{array}{c} \varepsilon_x \\ \gamma_{xy} \\ \kappa_x \end{array} \right\} = \left[\begin{array}{ccc} \dfrac{1}{A_{11}} & 0 & 0 \\ 0 & \dfrac{1}{A_{66}} & 0 \\ 0 & 0 & \dfrac{1}{D_{11}} \end{array} \right] \left\{ \begin{array}{c} N_x \\ N_{xy} \\ M_x \end{array} \right\} \qquad (5.8)$$

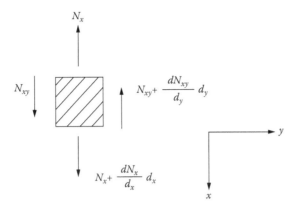

FIGURE 5.28
Isolated panel elements.

Based on force equilibrium for an infinitesimal section of the panel as shown in Figure 5.28, the equilibrium equations can be expressed as

$$\frac{\partial N_x}{\partial x} + \frac{\partial N_{xy}}{\partial y} = 0 \tag{5.9}$$

The compatibility equation is given as

$$\frac{\partial^2 \varepsilon_x}{\partial y^2} + \frac{\partial^2 \varepsilon_y}{\partial x^2} = \frac{\partial^2 \gamma_{xy}}{\partial x \partial y} \tag{5.10}$$

Neglecting transverse normal strain and assuming that remains constant along the y direction, the governing differential equation can be obtained by substituting (5.9) and into (5.8) as

$$\frac{1}{A_{11}} \frac{\partial^2 N_x}{\partial y^2} + \frac{1}{A_{66}} \frac{\partial^2 N_{xy}}{\partial x^2} = 0 \tag{5.11}$$

Equation (5.11) can be reduced to an ordinary differential equation by using harmonic analysis proposed by Salim and Davalos (2005), which was used to analyze shear lag for thin-walled open and closed composite beams. The panel in Figure 5.25 is simply supported at $x = 0, a$. Thus, the axial panel force can be define as

$$N_x(x,y) = \sum_{j=1}^{\infty} N_j(y) \sin(\frac{j\pi x}{a}) \tag{5.12}$$

where $N_j(y)$ is an amplitude function. Substituting (5.12) into (5.11) leads to

$$\frac{\partial^2 N_j}{\partial y^2} = \xi_j^2 N_j, \quad (\xi_j = \frac{j\pi}{a}\sqrt{\frac{A_{11}}{A_{66}}})$$

(5.13)

The general solution for (5.13) is given as

$$N_j(y) = C_{1j}\cosh(\xi_j y) + C_{2j}\sinh(\xi_j y)$$

(5.14)

where C_{1j} and C_{2j} are coefficients that can be determined by boundary conditions and loading conditions at the stiffened edges of the panel. Therefore, the variation of shear flow can be expressed as

$$\frac{\partial N_{xy}}{\partial y} = -\sum_{j=1}^{\infty}\frac{j\pi}{a}[C_{1j}\cosh(\xi_j y) + C_{2j}\sinh(\xi_j y)]\cos(\frac{j\pi x}{a})$$

(5.15)

The in-plane shear variation can be defined as (Barbero et al. 1993)

$$\frac{\partial N_{xy}}{\partial y} = -\frac{V(x)}{D}[\bar{A}e(y) + \bar{B}\cos\varphi]$$

(5.16)

where $\bar{A}$ is the extensional stiffness of the cross section, $\bar{B}$ is the bending-extension coupling stiffness, which can be neglected since the orthotropic FRP panel is usually designed as symmetric and balanced, $e(y)$ is the distance between the neutral axis of the cross section and the middle surface of the flange, $V(x)$ is the resultant shear force acting on the cross section, D is the cross section bending stiffness, and φ is the orientation of nonhorizontal flange. The in-plane variation of shear

$$\frac{\partial N_{xy}}{\partial y}$$

can be written in the form of Fourier series as

$$\frac{\partial N_{xy}}{\partial y} = \sum_{j=1}^{\infty}Q_j\cos(\frac{j\pi x}{a})$$

(5.17)

And the coefficient in (5.17) can be defined as

$$Q_j = \frac{2}{a}\int_0^a\frac{\partial N_{xy}}{\partial y}\cos(\frac{j\pi x}{a})dx$$

(5.18)

Substituting (5.16) into (5.18), we have

$$Q_j = -\frac{2\overline{A}e(y)}{aD}I_j \qquad (5.19)$$

where I_j depends on loading condition. If the origin of the y axis is located at the center of the cross section, then $C_{2j} = 0$ due to symmetry, and (5.14) can be reduced to

$$N_j(y) = C_{1j}\cosh(\xi_j y) \qquad (5.20)$$

By ensuring compatibility of shear flow at the junction of flange and web ($y = -b/2, b/2$), C_{1j} can be obtained by equating (5.15) and (5.17). Therefore, the normal force resultant and normal stress along the panel can be obtained as

$$N_x(x,y) = -\sum_{j=1}^{\infty}\frac{a}{j\pi}Q_j[\frac{\cosh(\xi_j y)}{\cosh(\frac{\xi_j b}{2})}]\sin(\frac{j\pi x}{a}) \qquad (5.21)$$

$$\sigma_x(x,y) = \frac{N_x(x,y)}{\overline{A}} \qquad (5.22)$$

Based on definition of effective flange width, the longitudinal normal stress is assumed to be uniformly distributed along the panel section, as shown in Figure 5.2b. The effective flange width can be expressed as the integral of normal stress distribution divided by the maximum stress σ_{max} at the panel-stiffener intersection as

$$b_e = \frac{\int_{-b/2}^{b/2}\sigma_x \, dx}{\sigma_{max}} = \frac{2\int_{0}^{b/2}\sigma_x \, dx}{\sigma_{max}} \qquad (5.23)$$

Finally, substituting (5.21) and (5.22) into (5.23), and taking only the first-term approximation for simplicity, the expression for effective flange width is given as

$$b_e = \frac{\int_{0}^{b/2}\cosh(\xi_1 y)\,dy}{\cosh(\frac{\xi_1 b}{2})} \qquad (5.24)$$

5.3.3 Finite Element Study

5.3.3.1 FE Model Descriptions

To verify the shear lag model developed in the previous section, FE models for 44 simple-span FRP deck-over-steel girder bridges were constructed using ABAQUS (2002) (Zou et al. 2011). The bridge deck was an FRP honeycomb panel produced by Kansas Structural Composites, Inc. (KSCI), which can be idealized as a structurally orthotropic panel with homogeneous equivalent engineering properties, as shown in Table 5.15 (Davalos et al. 2001). The bridge configurations considered are shown in Figure 5.29 and Table 5.16. Two and three lanes were considered, with widths of 9.31 and 12.97 m, respectively. Other varied geometric parameters included span length, with 11 lengths selected from 15.24 to 91.44 m at increments of 7.62 m; and girder spacing, with 1.98, 2.59, and 3.51 m for two-lane bridges, and 3.51 m for three-lane bridges. The FRP panel and steel I-beam components were modeled using shell elements (S4R), and beam elements were used to model cross-frame bracings. Multiple-point constraint (MPC) rigid elements were used to simulate the interaction between the panel and girders.

The following assumptions are adopted in the FE study to simplify the analysis effort while retaining adequate accuracy: (1) The bridge deck is idealized as a homogeneous, elastic, and orthotropic slab with uniform thickness. (2) The slab is supported by equally spaced I-shaped steel girders. (3) The edges of the slab-and-girder ends are simply supported at the abutments. (4) Full composite action is assumed between the supporting girders

TABLE 5.15

Equivalent Properties of FRP Panel

	E_x (MPa)	E_y (MPa)	v_x	G_{xy} (MPa)
In-plane	2,560	2,300	0.303	560
Bending	5,640	5,640	0.303	1,400

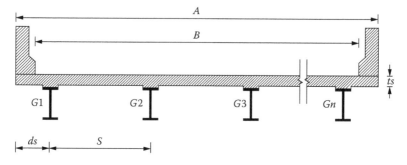

FIGURE 5.29
Typical cross section of bridge model.

TABLE 5.16

Parameter for Each Cross Section[*]

Parameter	CS1	CS2	CS3	CS4
A	9.3 m	9.3 m	9.3 m	13.0 m
B	8.53 m	8.53 m	8.53 m	12.2 m
S	3.51 m	2.60 m	1.98 m	3.51 m
ds	1.15 m	0.77 m	0.69 m	1.23 m
ts	0.24 m	0.20 m	0.20 m	0.24 m
N	3	4	5	4

[*] See Figure 5.29 for definition of parameters.

and slab, i.e., no interface slip at the girder-slab interface. (5) Based on design guidelines, no truck wheel load can be placed closer than 0.61 m from the roadway edge. For simplicity, the assumption of full composite action is adopted in the FE modeling, with the purpose of verifying the accuracy of the shear lag model developed in the previous section. In general, however, and as noted by Machado et al. (2008), most FRP deck-on-steel girder systems are characterized by partial composite action. In that case, a reduction factor may be applied to full composite action given by the present shear lag model in order to estimate effective flange width for partial composite action, as will be shown later.

5.3.3.2 Live Load Position

The AASHTO HS20 truckload from AASHTO LRFD specifications (2004) was adopted. It was positioned longitudinally at a selected location to induce maximum moment in the bridge models. For three cross-sectional bridge cases, one- and two-lane load conditions were selected, and for one cross-sectional case, one-, two-, and three-lane load conditions were evaluated, as shown in Figure 5.30.

5.3.3.3 Data Reduction from FE Results

Similar to the shear lag model, (5.23) was used to calculate effective flange width. The average normal stress, σ_x, which can be assumed uniform along its segmentally discrete width h, can be obtained from the output for each shell element, as shown in Figure 5.31. The integral of the normal stresses, or normal stress resultant, can be approximated as the summation of the discrete values over the panel section, which can be expressed as

$$\int_{-b/2}^{b/2} \sigma_x \, dx = \sum_{i=1}^{n} \sigma_{x,i} h_i \tag{5.25}$$

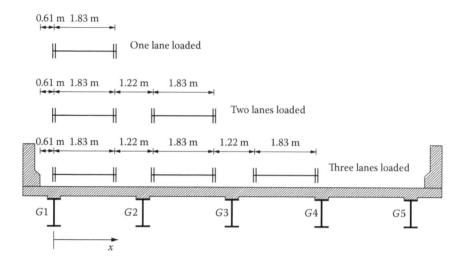

FIGURE 5.30
AASHTO HS20 truck live load.

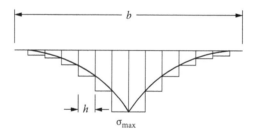

FIGURE 5.31
Stress integration along flange width.

where $\sigma_{x,i}$ is the normal stress for the shell element i, and n is the total number of elements. In (5.23), σ_{max} is the stress value at the location of the panel-girder intersection.

5.3.3.4 Comparison between Shear Lag Model and FE Analysis

Effective flange widths from both shear lag model and FE analysis are shown in Table 5.17. Both results show similar trends; i.e., the effective width increases as the spacing-span aspect ratio, S/L, decreases. When the aspect ratio is less than 0.1, the effective width is close to 96% of flange width. The largest difference is found for the model with 7.62 m span length, which is about 8 to 14%. Overall, the effective flange widths predicted from the shear lag model are close to the FE results, and the shear lag model overestimates the effective flange width by an average of 6%.

TABLE 5.17

Comparison of Shear Lag Model and FE

Cross Section	Span (m)	S/L	b_e/S Shear Lag	B_e/S FE	Difference (%)
CS1	7.62	0.460	0.626	0.551	11.997
	38.1	0.092	0.974	0.912	6.354
	68.6	0.051	0.992	0.941	5.127
	91.4	0.038	0.995	0.930	6.541
CS2	7.62	0.340	0.744	0.642	13.667
	38.1	0.068	0.986	0.925	6.179
	68.6	0.038	0.995	0.962	3.315
	91.4	0.028	0.997	0.923	7.502
CS3	7.62	0.260	0.829	0.762	8.004
	38.1	0.052	0.991	0.966	2.525
	68.6	0.029	0.997	0.975	2.209
	91.4	0.022	0.999	0.911	8.774
CS4	7.62	0.460	0.626	0.537	14.172
	38.1	0.092	0.974	0.892	8.436
	68.6	0.051	0.992	0.955	3.749
	91.4	0.038	0.995	0.935	6.023

5.3.4 Comparison between Shear Lag Model and Empirical Functions

Tenchev (1996) conducted a parametric FE study on effective width for orthotropic plates considering pin-roller and fixed-fixed boundary conditions, concentrated and uniformly distributed loading conditions, and different cross sections. A total of 640 FE models were analyzed to obtain the effective flange width, based on which empirical functions were proposed using the regression technique as

$$\lambda_{emp} = \frac{b_e}{b} = \frac{0.57}{C_1 C_2} \left(\frac{b}{L}\right)^{-0.85} \left(\frac{E}{G}\right)^{-0.416} \tag{5.26}$$

$$C_1 = 1 + 5e^X, \; X = -6.4 \frac{b}{L} \sqrt{\frac{E}{G}} \tag{5.27}$$

$$C_2 = 1 + 0.31e^{-0.9Y}, \; Y = \left(\frac{E}{G}\right)\left(\frac{b}{L}\right)^{-1} \tag{5.28}$$

where b_e = effective flange width, b = actual beam flange width, L = span length, E = beam flange Young's modulus in longitudinal direction, and G = beam flange in-plane shear modulus.

Both the proposed shear lag model and the empirical function of (5.26) were used to study a simulated T-beam model with varying modulus ratio E/G and aspect ratio b/L. Three modulus ratios were selected: 1, 10, and 30. The aspect ratio was varied from 0.1 to 1 for $E/G = 1$, from 0.08 to 0.88 for $E/G = 10$, and finally from 0.03 to 0.60 for $E/G = 30$. The boundary condition was assumed to be simply supported and the loading was assumed to be uniformly distributed.

The results are compared in Tables 5.18 to 5.20 for $E/G = 1$, 10, and 30, respectively. For $E/G = 1$, the shear lag model consistently overestimates the

TABLE 5.18

Comparison between Shear Lag Model and Empirical Function for $E/G = 1$

	b_e/b		
b/L	Empirical	Shear Lag	Difference (%)
1.00	0.502	0.584	16.29
0.90	0.551	0.628	14.06
0.80	0.608	0.676	11.29
0.70	0.673	0.728	8.19
0.60	0.743	0.781	5.13
0.50	0.812	0.835	2.85
0.40	0.867	0.886	2.18
0.30	0.901	0.932	3.42
0.20	0.933	0.968	3.75
0.10	1.000	0.992	−0.81
		Average	6.63

TABLE 5.19

Comparison between Shear Lag Model and Empirical Function for $E/G = 10$

	b_e/b		
b/L	Empirical	Shear Lag	Difference (%)
0.88	0.244	0.229	−6.20
0.78	0.270	0.258	−4.58
0.69	0.300	0.291	−2.91
0.60	0.338	0.334	−1.15
0.52	0.381	0.383	0.38
0.43	0.448	0.455	1.67
0.34	0.544	0.553	1.58
0.26	0.670	0.665	−0.67
0.17	0.850	0.815	−4.13
0.08	0.940	0.950	1.08
		Average	−1.49

TABLE 5.20

Comparison between Shear Lag Model and Empirical
Function for $E/G = 30$

b/L	b_e/b Empirical	b_e/b Shear Lag	Difference (%)
0.60	0.214	0.193	−9.53
0.54	0.234	0.215	−8.09
0.48	0.258	0.242	−6.48
0.41	0.295	0.282	−4.51
0.35	0.338	0.330	−2.37
0.29	0.397	0.395	−0.42
0.22	0.500	0.504	0.77
0.16	0.646	0.639	−1.06
0.10	0.852	0.809	−5.08
0.03	0.993	0.978	−1.49
		Average	−3.83

effective width. The average difference is about 6.6% between the two models. The results are very close for $b/L < 0.5$, as shown in Table 5.18. For $E/G = 10$ and 30, the average differences are 1.5 and 3.8%, respectively. Overall, there is a good correlation between the two models, which can further verify the accuracy of the shear lag model developed in Section 5.3.2. It is noted that, because of the nature of the regression technique, the applications of the empirical functions are limited, while the present shear lag model can be applied to a wider range of problems and is more suitable for design and parametric study purposes.

5.3.5 Application of Shear Lag Model to FRP Deck

The proposed shear lag model was further used to calculate effective flange width for the scaled FRP deck-on-steel girder bridge model as shown in Section 5.2. Using the shear lag model, the effective flange width is predicted to be 1.01 m, which is higher than the testing results because full composite action is assumed in the present model. Therefore, a reduction factor R is suggested in order to account for the effect of partial composite action. Accordingly, based on (5.24), we can define b_e, for partial composite action, as

$$b_e = R \frac{\displaystyle\int_0^{b/2} \cosh(\xi y)\, dy}{\cosh\left(\dfrac{\xi b}{2}\right)}, \quad \xi = \frac{\pi}{a}\left(\sqrt{\frac{A_{11}}{A_{66}}}\right) \tag{5.29}$$

For the scaled bridge model described above, $R = 0.62$ based on a correlation between testing and analytical results, corresponding to a 25% composite action as reported in Section 5.2. Thus, based on an approximate regression of limited data, the following equation is proposed to calculate R for a given DCA as

$$R = 1.025(1 - 0.0244^{DCA})$$ (5.30)

which is shown in Figure 5.32.

5.3.6 Conclusions

In this section, a shear lag model is proposed to calculate effective flange width for orthotropic decks on steel girder bridges. To verify this solution, a finite element parametric study is conducted on 44 simply supported FRP deck-on-steel girder bridges assuming full composite action. By comparing effective flange widths from FE analysis and a shear lag model, it is found that the shear lag model predicts the effective flange width fairly well for interior girders, with an average difference of about 6%. The accuracy of this model is further verified by favorable correlation with an existing empirical solution, with average differences of 6.6, 1.5, and 3.8% for $E/G = 1$, 20, and 30, respectively.

By comparison between the analytical and testing results for the T-beam model shown in Section 5.2, it is illustrated that the proposed shear lag model can be further applied to predict effective flange width for bridges

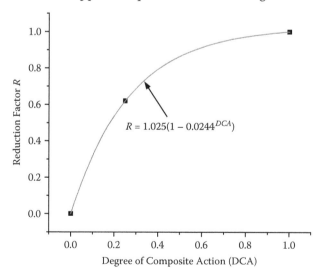

FIGURE 5.32
Reduction factor *R*–DCA curve.

with partial DCA, with the introduction of a reduction factor R, for a specified percent of composite action. To this effect, an empirical relation between DCA and R is suggested.

References

AASHTO. (1996). *AASHTO standard specifications for highway bridges.* 16th ed. American Association of State Highway and Transportation Officials, Washington, DC.

AASHTO. (2004). *AASHTO LRFD bridges design specifications.* 3rd ed. American Association of State Highway and Transportation Officials, Washington, DC.

ABAQUS. (2002). *User's manual.* Version 6.3. HKS, Providence, RI.

Adekola, A.O. (1968). Effective widths of composite beams of steel and concrete. *Structural Engineer,* 46(9), 285–289.

Adekola, A.O. (1974). The dependence of shear lag on partial interaction in composite beams. *International Journal of Solids Structures,* 10, 389–400.

Ahn, I., Chiewanichakorn, M., Chen, S.S., and Aref, A.J. (2004). Effective flange width provisions for composite steel bridges. *Engineering Structure,* 26, 1843–1851.

Amadio, C., Fedrigo, C., Fragiacomo, M., and Macorini, L. (2004). Experimental evaluation of effective width in steel-concrete composite beams. *Journal of Constructional Steel Research,* 60, 199–220.

Bakis, C.E., Bank, L.C., Brown, V.L., Cosenza, E., Davalos, J.F., Lesko, J.J., Machida, A., Rizkalla, S.H., and Triantafillou, T.C. (2002). Fiber-reinforced polymer composites for construction-state-of-art review. *ASCE Journal of Composites for Construction,* 6(2), 73–87.

Barbero, E.J., Lopez-Anido, R., and Davalos, J.F. (1993). On the mechanics of thin-walled laminated composite beams. *Journal of Composite Materials,* 27(8), 806–829.

Chen, A., and Davalos, J.F. (2010). Strength evaluations of sinusoidal core for FRP sandwich bridge deck panels. *Composite Structures,* 92(7), 1561–1573.

Cheung, M.S., and Chan, M.Y.T. (1978). Finite strip evaluation of effective flange width of bridge girders. *Canadian Journal of Civil Engineering,* 5(2), 174–185.

CSA. (2000). *Canadian highway bridge design code (CAN/CSA-S6-00).* CSA International.

Dassault Systèmes. (2007). *ABAQUS/CAE user's manual.* Version 6.7. Pawtucket, RI.

Davalos, J.F., and Chen, A. (2005). Buckling behavior of honeycomb FRP core with partially restrained loaded edges under out-of-plane compression. *Journal of Composite Materials,* 39(16), 1465–1485.

Davalos, J.F., Chen A., and Zou, B. (2011). Stiffness and strength evaluations of a new shear connection system for FRP bridge decks to steel girders. Accepted by *ASCE Journal of Composite for Construction,* 15(3), 441–450.

Davalos, J.F., Chen, A., and Zou, B. (2012). Experimental investigation of a scaled FRP deck-on-steel girder bridge model with partial degree of composite action. *Engineering Structures,* 40, 51–63.

Davalos, J.F., Qiao, P.Z., Xu, X.F., Robinson, J., and Barth, K.E. (2001). Modeling and characterization of fiber-reinforced plastic honeycomb sandwich panels for highway bridge applications. *Composite Structures,* 52, 441–452.

Davalos, J.F., and Salim, H.A. (1993). Effective flange width for stress-laminated T-system timber bridges. *Journal of Structural Engineering*, 119(3), 938–953.

Demitz, J.R., Mertz, D.R., and Gillespie, J.W. (2003). Deflection requirements for bridges constructed with advanced composite materials. *Journal of Bridge Engineering*, 8(2), 73–83.

Eom, J., and Nowak, A.S. (2001). Live load distribution for steel girder bridges. *Journal of Bridge Engineering*, 6(6), 489–497.

Keelor, D.C., Luo, Y., Earls, C.J., and Yulismana, W. (2004). Service load effective compression flange width in fiber reinforced polymer deck systems acting compositely with steel stringers. *Journal of Composites for Construction*, 8(4), 289–297.

Keller, T., and Gürtler, H. (2005). Composite action and adhesive bond between fiber-reinforced polymer bridge decks and main girders. *Journal of Composites for Construction*, 9(4), 360–368.

Keller, T., and Tirelli T. (2004). Fatigue behavior of adhesively connected pultruded GFRP profiles. *Composite Structures*, 65, 55–64.

Machado, A.S.M., Sotelino, E.D., and Liu, J. (2008). Modeling technique for honeycomb FRP deck bridges via finite elements. *ASCE Journal of Structural Engineering*, 134(4), 572–580.

Moffat, K.R., and Dowling, P.J. (1975). Shear lag in steel box girder bridges. *Structural Engineer*, 53(10), 439–448.

Moffat, K.R., and Dowling, P.J. (1978). British shear lag rules for composite girders. *Journal of the Structural Division*, 104(7), 1123–1130.

Moon, F.L., and Gillespie, J.W. (2005). Experimental validation of a shear stud connection between steel girders and a fiber-reinforced polymer deck in the transverse direction. *Journal of Composites for Construction*, 9(3), 284–287.

Park, K.T., Kim, S.H., Lee, Y.H., and Hwang Y.K. (2005). Degree of composite action verification of bolted GFRP bridge deck-to-girder connection system. *Composite Structures*, 72, 393–400.

Salim, H.A., and Davalos, J.F. (2005). Shear lag of open and closed thin-walled laminated composite beams. *Journal of Reinforced Plastics and Composites*, 24(7), 673–690.

Song, Q., and Scordelis, A.C. (1990a). Shear-lag analysis of T-, I-, and box beams. *Journal of Structural Engineering*, 116(5), 1290–1305.

Song, Q., and Scordelis, A.C. (1990b). Formulas for shear-lag effect of T-, I-, and box beams. *Journal of Structural Engineering*, 116(5), 1306–1319.

Tenchev, R.T. (1996). Shear lag in orthotropic beam flanges and plates with stiffeners. *International Journal of Solids Structures*, 33(9), 1317–1334.

Zhang, Y., and Cai, C.S. (2007). Load distribution and dynamic response of multi-girder bridges with FRP decks. *Engineering Structures*, 29(8), 1676–1689.

Zou, B. (2008). Design guidelines for FRP honeycomb sandwich bridge decks. PhD dissertation, West Virginia University, Morgantown.

Zou, B., Chen, A., Davalos, J.F., and Salim, H. A. (2011). Evaluation of effective flange width by shear lag model for orthotropic FRP bridge decks. *Composite Structures*, 93, 474–482.

6

Design Guidelines for FRP Deck–
Steel Girder Bridge Systems

Based on research findings presented in previous chapters and Chen (2004), design guidelines are proposed in this chapter for all three bridge components: fiber-reinforced polymer (FRP) deck, shear connector, and bridge system. A design example is provided to illustrate the use of the proposed design guidelines. While the recommendations given for FRP decks are focused on a specific sandwich panel, the guidelines for a shear connector and bridge system can be used for other types of FRP decks.

6.1 Design Guidelines

6.1.1 FRP Deck

As shown in Figure 6.1, core materials for sandwich structures are primarily subjected to out-of-plane compression and shear, and the facesheet laminates sustain mainly membrane forces due to bending. As pointed out in Chapter 3, pure compression and elastic buckling are two failure modes for out-of-plane compression. Shear crushing, shear buckling, and delamination can occur for out-of-plane shear. The facesheet and core of the honeycomb fiber-reinforced polymer (HFRP) sandwich panels are attached by contact molding, and are therefore not rigidly connected. Thus, the buckling of the core can be described as the instability of an FRP core panel with two rotationally restrained edges, where the degree of the restraint is dependent on the bonding layer thickness between the core and facesheet. Therefore, the core strength is controlled by two parameters: bonding layer effect and core thickness (t, as shown in Figure 6.3). The strength of the facesheet is dependent on the configuration of the lay-ups. As shown in Chapter 3, three bonding layers and three core thicknesses are considered, with the naming convention defined in Figure 3.6, where the letters B and C represent, respectively, chopped strand mat (ChSM) bonding layer numbers, and core thicknesses. The integers i and j ($i = j = 1, 2, 3$) correspond to their respective nominal weights of ChSM layer used. Three different types of facesheet laminates are considered, as shown in Table 6.1. Design guidelines are proposed considering various failure modes as follows.

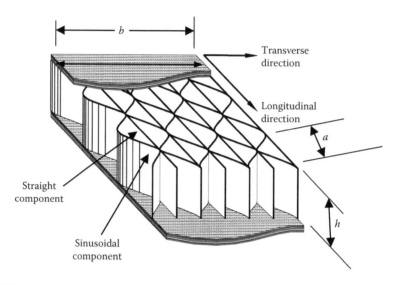

FIGURE 6.1
HFRP panel with sinusoidal core configuration.

TABLE 6.1

Plate Configurations for Facesheet Laminate

	Laminate 1	Laminate 2	Laminate 3
	2 layers 900 g/m² ChSM	2 layers 900 g/m² ChSM	2 layers 900 g/m² ChSM
	1 layer biaxial	1 layer biaxial	8 layers biaxial
	9 layers uniaxial	1 layer 900 g/m² ChSM	
	1 layer biaxial	1 layer biaxial	
		1 layer 900 g/m² ChSM	
		1 layer biaxial	
		1 layer 900 g/m² ChSM	
		1 layer biaxial	
		1 layer 900 g/m² ChSM	
Thickness	15.0 mm (0.59 in.)	16.5 mm (0.65 in.)	14.2 mm (0.56 in.)

Note: Biaxial: CDM 3208.
    Uniaxial: CM 1708.
    ChSM: Chopped strand mat.

6.1.1.1 Out-of-Plane Compression

Two distinct failure modes may occur for a panel under out-of-plane compression: pure compression and elastic buckling, where the final failure load depends on the lowest value from these failure modes. Figure 6.2 can be used to predict compression failure load. The following method is proposed:

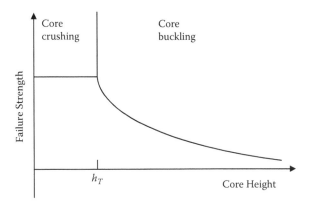

FIGURE 6.2
Design diagram for compression and shear.

1. In relation to the interface bonding layer thickness (number of layers), a transition height h_T is defined, above which the core undergoes buckling failure. Thus, we compare the height of the panel h with the transition height h_T, as shown in Table 6.2. If $h < h_T$, the failure mode is pure compression failure, and we use material compressive strength f_c, which can be obtained from coupon test results in Chapter 3, as controlling strength; otherwise, buckling dominates the failure, and we evaluate strength according to

$$\sigma = 4.4482 \times \frac{A_1 e^{[-h/(B_1 \times 25.4)]} + A_2 e^{[-h/(B_2 \times 25.4)]} + F_0}{A} \tag{6.1}$$

where h is the height of the panel, and all the other parameters are listed in Table 6.3.

2. We calculate the compressive stress based on the most critical loading condition as

$$\sigma_c = \frac{F}{A_c} \tag{6.2}$$

where F is the out-of-plane compression force, and A_c is the total in-plane area of the core walls (Figure 6.3). This stress can be compared with the compressive strength obtained from step 1 to find the safety factor.

TABLE 6.2

Transition Height for Compression

	One Bonding Layer	Two Bonding Layers	Three Bonding Layers
h_T	32 mm	36 mm	38 mm

TABLE 6.3

Parameters for Equation (6.1)

	A_1	B_1	A_2	B_2	F_0
One bonding layer	957,515	0.2363	124,742	0.7464	8,081
Two bonding layers	87,639	1.0105	954,711	0.2917	8,136
Three bonding layers	1,038,189	0.2985	88,384	1.0765	8,152

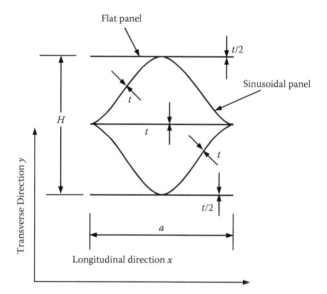

FIGURE 6.3
Unit cell dimension.

6.1.1.2 Out-of-Plane Shear

Three distinct failure modes may occur for a panel under out-of-plane shear: shear crushing and shear buckling for the flat component, and interface debonding for the sinusoidal component, where the final failure load depends on the lowest value from these failure modes. Figure 6.2 can be schematically used to define failure strength for the flat core component. The following method is proposed:

1. We compare the height of the panel h with the transition height h_T, as shown in Table 6.4. If $h < h_T$, the failure mode is pure shear failure, and we use the material shear strength as a controlling strength; otherwise, buckling dominates the failure, and we define strength as

$$\tau = 0.175 \times (\frac{A_1}{t} e^{-\frac{R}{B_1}} + \frac{A_2}{t} e^{-\frac{R}{B_2}} + \frac{N_0}{t}) \tag{6.3}$$

where $R = h/a$ is the aspect ratio, t is the thickness of the core wall, and all other parameters are given in Table 6.5. The debonding needs to be evaluated separately in step 3.

2. Equivalent shear modulus G_{xz} can be calculated based on the homogenization theory proposed in Chapter 2. For two core thicknesses (C2), a more accurate G_{xz} can be calculated based on Chapter 3 considering skin effect as

$$G_{xz} = 313.88 + 5.23e^{\frac{R - 0.1113}{-0.7987}} + 0.99e^{-\frac{R - 0.1113}{20}} \tag{6.4}$$

where R is the aspect ratio. Next, we calculate shear strain based on induced shear load per unit transverse core area:

$$\gamma = \frac{V}{G_{xz}bh} \tag{6.5}$$

where b and h are the width and height of the cross section, respectively, as shown in Figure 6.1, and G_{xz} is the equivalent shear stiffness from (6.4). It is noted that the shear strain for the flat core component is the same as the global shear strain. Therefore, the shear stress in the flat core component is

$$\tau_{12} = G_{12}\gamma \tag{6.6}$$

where G_{12} is the material shear stiffness. And we compare this stress with the shear strength obtained from step 1 to get the safety factor.

3. Using shear strain obtained from step 2, we can find the interfacial tensile stress for the curved core component based on warping

TABLE 6.4

Transition Height for Shear

	One Bonding Layer	Two Bonding Layers	Three Bonding Layers
h_T	88 mm	94 mm	98 mm

TABLE 6.5

Parameters for Equation (6.3)

	A_1	B_1	A_2	B_2	N_0
One bonding layer	2,103	0.5326	34,611	0.1388	448
Two bonding layers	2,661	0.5097	37,093	0.1355	449
Three bonding layers	3,015	0.4970	38,734	0.1339	450

theory, as shown in Chapter 3. For two core thicknesses (C2), the interfacial tensile stress can be calculated as

$$\sigma_{interface} = \gamma(1655.78 - 1655.78e^{-[0.09123h)]^{1.0221}})$$ (6.7)

where γ is the shear strain, and h is the height of the panel. We then compare the interfacial tensile stress $\sigma_{interface}$ with the nominal interfacial tensile strength in Table 6.6 to define the safety factor.

6.1.1.3 Facesheet Check

The bending moment is carried through the membrane forces of the facesheet, as shown in Figure 6.4. And the compressive and tensile force can be calculated as

$$C = T = \frac{M}{h}$$ (6.8)

Compressive force usually controls the design. This force can be compared with experimental results as given in Table 6.12 to find the safety factor.

6.1.2 Shear Connector

The prototype shear stud sleeve-type connection as described in Chapter 4 will be adopted. To design for the shear connector, the interfacial shear for each connector needs to be calculated and compared with the strength obtained experimentally in Chapter 4, as described below.

TABLE 6.6

Nominal Interfacial Tensile Strength

	B1C2	B2C2	B3C2	B2C1	B2C3	B3C1	B3C3
Nominal interfacial tensile strength (MPa)	8.3	12.1	15.1	8.8	11.8	11.8	20.7

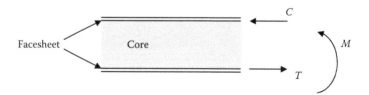

FIGURE 6.4
Forces acting on facesheet.

6.1.2.1 Static Strength and Load-Displacement Formulation (P-Δ Curve)

For the sleeve-type connector as shown in Figure 4.2, the load-displacement curves are segmentally linear, as shown in Figure 4.9, with three ranges that can be described as

$$\begin{cases} k = 1.5 \text{ kN/mm}, & \Delta = 0\text{–}15 \text{ mm} \\ k = 7.9 \text{ kN/mm}, & \Delta = 15\text{–}25 \text{ mm} \\ k = 1.4 \text{ kN/mm}, & \Delta > 25 \text{ mm} \end{cases}$$

where k is the stiffness of the connection. The yield strength is 102 kN and the ultimate strength is 120 kN.

6.1.2.2 Fatigue Strength and S-N Curve

The fatigue test results are shown in Figure 4.10 in terms of fatigue load cycles and stress ranges, which can be fitted by a logarithm function as

$$\log N = 7.6 - 0.01096S \tag{6.9}$$

where N is number of load cycles and S is stress range of the shear connection in MPa.

6.1.3 Bridge System

6.1.3.1 Effective Flange Width

Based on findings from Chapter 5, the effectively flange width b_{eff} for the honeycomb FRP sandwich deck section connected to a steel girder, using the mechanical shear connector with a partial composite action, can be calculated as

$$b_{eff} = R \frac{\displaystyle\int_0^{b/2} \cosh(\xi y)\,dy}{\cosh\left(\dfrac{\xi b}{2}\right)}, \quad \xi = \frac{\pi}{a}\left(\sqrt{\frac{A_{11}}{A_{66}}}\right) \tag{6.10}$$

R is a reduction factor as

$$R = 1.025(1 - 0.0244^{DCA}) \tag{6.11}$$

where DCA = degree of composite action. For design purposes, (6.10) can be simplified as

$$b_{eff} = b_{eff\,AASHTO} \times R \qquad\qquad (6.12)$$

where $b_{eff\,AASHTO}$ is the effective flange width calculated from AASHTO specifications. As pointed out in Chapter 5, (6.12) provides a conservative result.

6.1.3.2 Load Distribution Factor

As shown in Chapter 5, the load distraction factor can be calculated based on AASHTO LRFD specifications.

6.2 Example

An example is provided to illustrate the use of the proposed design guidelines. The design is based on AASHTO LRFD specifications (2007). The configuration of the example bridge (Figure 6.5) and design assumptions are described below:

1. Simply supported 21.33 m span.
2. 10.97 m wide accommodating two design lanes.
3. Five W40 × 199 grade 50 rolled steel girders at 2.44 m on centers, with a yield strength $f_y = 345$ MPa.
4. 254 mm thick FRP honeycomb sandwich deck panel connected to steel girders, using the shear connection as described in Chapter 4. Assume the spacing is 1.22 m on centers and a DCA of 25%.
5. The deck configuration is shown in Figure 6.6. The properties of facesheet and core material are listed in Table 6.7 through Table 6.9. Assume using Laminate 1L as facesheet, one bonding layer, and two core thicknesses (B1C2).

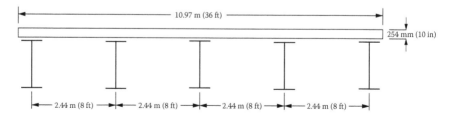

FIGURE 6.5
Cross section of the example bridge.

TABLE 6.7

Material Properties of Facesheet

		Nominal Weight (g/m²)	Thickness (mm)	V_f
CM 3205	0° or 90°	542.5	0.620	0.3428
	CSM	152.6	0.254	0.2359
UM 1810	0°	610.3	0.635	0.3774
	CSM	305.2	0.335	0.3582
Bonding layer	ChSM	600	3.175	0.1726

TABLE 6.8

Stiffness Properties of Facesheet Lamina

	Orientation	E_1 (GPa)	E_2 (GPa)	G_{12} (GPa)	G_{23} (GPa)	v_{12}	v_{22}
CM 3205	0 or 90°	27.72	8.00	3.08	2.88	0.295	0.390
	Random	11.79	11.79	4.21	2.36	0.402	0.400
UM 1810	0°	30.06	8.55	3.30	3.08	0.293	0.386
	Random	15.93	15.93	5.65	2.96	0.409	0.388
Bonding layer	Random	9.72	9.72	3.50	2.12	0.394	0.401

TABLE 6.9

Stiffness Properties of Facesheet and Core

	E_x (GPa)	E_y (GPa)	v_x	G_{xy} (GPa)
Facesheet	19.3	12.35	0.32	3.812
Core	0.529	0.000986	0.431	0.000705

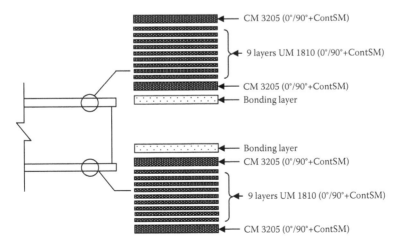

FIGURE 6.6
FRP deck configurations.

6. Compressive strength of the facesheet is assumed to be 50% of the compressive strength from Table 6.12 for Laminate 1L, i.e., 2.33 kN/mm (13.4 kips/in.).

7. For simplicity, calculations are only provided for an interior girder under flexural conditions to meet the requirements of strength, service; and shear connector to meet the requirements of strength.

6.2.1 FRP Deck

Using the design guidelines provided in the previous section, we can evaluate the strength of an HFRP sandwich panel for a bridge under a patch load. The height of the panel $h = 254$ mm, the core wall thickness for both curved and flat is $t = 2.3$ mm, the length of the component $a = 102$ mm, and the aspect ratio $R = h/a = 2.5$. One bonding layer is used at the interface between the core and facesheet. The HFRP panel can be treated as a one-way slab supported by floor beams, with two sides simply supported and two sides free, as shown in Figure 6.7. The maximum applied load corresponds to AASHTO HS20-44 (AASHTO 2007) design truck wheel load (71,171 N) with a dynamic load allowance of 33%:

$$P_{max} = 71,171 \times 1.33 = 94,658 \ N \tag{6.13}$$

According to AASHTO LRFD specifications (2007), the width of the contact area between the wheel and bridge deck is 508 mm (20 in.), and the length can be calculated as (AASHTO 2007, Section 3.6.1.2.5):

$$\ell = 6.4\gamma(1 + IM/100) = 6.4(1 + 33/100) = 8.512 \ \text{in.} = 216 \ \text{mm} \tag{6.14}$$

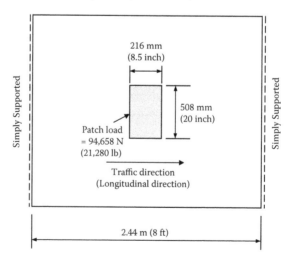

FIGURE 6.7
Panel layout.

where γ, the load factor, is assumed to be 1.0 for safety reasons; IM = dynamic allowance percent. It is assumed that the wheel load is evenly distributed over the contact area. It is worth pointing out that when designing a bridge deck, further distribution along the longitudinal direction is allowed (AASHTO 2007, Section 4.6.2), which is not considered because the purpose of this example is to illustrate how to evaluate the strength of the panel for a given loading condition. Proper code/specifications shall be followed in order to calculate the design forces, which is beyond the scope of this chapter.

6.2.1.1 Compressive Strength

The compressive stress can be calculated as

$$\sigma_c = \frac{P_{max}}{216 \times 508} = \frac{94,658}{216 \times 508} = 0.86 \; MPa \tag{6.15}$$

The panel height is 254 mm. From Table 6.12, we find that the transition height for one bonding layer is 32 mm, which is less than 254 mm, and therefore buckling controls the design. Substituting all the values into (6.1), the buckling strength is found to be 3.48 MPa, which gives a safety factor of 3.48/0.86 = 4.0. The calculation can be repeated for other core heights, and the results are given in Table 6.10.

6.2.1.2 Shear Strength

For convenience, we define a 25.4 mm wide beam with a distributed load q acting at mid-span, i.e., b = 25.4 mm, as shown in Figure 6.8. The distributed load q can be calculated as

$$q = \frac{P_{max}}{216 \times 508} = \frac{94,658 \times 25.4}{216 \times 508} = 21.9 \; N \; / \; mm \tag{6.16}$$

TABLE 6.10

Compressive Strength Check

Core Height (mm)	Pure Compressive Strength (MPa)	Buckling Strength (MPa)	Controlling Strength (MPa)	Safety Factor
13	15.9	80.7	15.9	18.5
51	15.9	7.3	7.3	8.4
102	15.9	3.7	3.7	4.3
203	15.9	3.5	3.5	4.0
254	15.9	3.5	3.5	4.0
305	15.9	3.5	3.5	4.0
508	15.9	3.5	3.5	4.0
1016	15.9	3.5	3.5	4.0

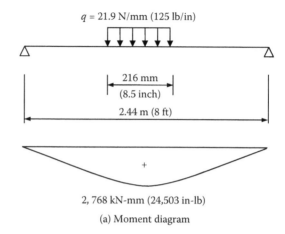

$q = 21.9$ N/mm (125 lb/in)

216 mm
(8.5 inch)

2.44 m (8 ft)

2, 768 kN-mm (24,503 in-lb)

(a) Moment diagram

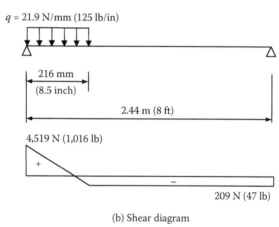

$q = 21.9$ N/mm (125 lb/in)

216 mm
(8.5 inch)

2.44 m (8 ft)

4,519 N (1,016 lb)

209 N (47 lb)

(b) Shear diagram

FIGURE 6.8
Unit width panel loading condition.

The maximum moment and shear can be obtained by placing the distributed load at the mid-span and also close to the support, respectively, as shown in Figure 6.8.

The height of the panel is 254 mm, and therefore the aspect ratio

$$R = \frac{254}{102} = 2.5$$

Using (6.4), we can find the equivalent shear stiffness $G_{xz} = 315$ MPa. Based on (6.5), the shear strain can be approximated as

$$\gamma = \frac{V}{G_{xz}bh} = \frac{4519}{315,000 \times 25.4 \times 254} = 0.00222 \qquad (6.17)$$

Therefore, according to (6.6), the shear stress in the flat core component is

$$\tau_{12} = G_{12}\gamma = 4,206 \times 0.00222 = 9.4 \ MPa \quad (1,354 \ psi) \tag{6.18}$$

where G_{12} (4,206 MPa) is the material shear modulus from material test data as shown in Chapter 3. From Table 6.3, we find $h_T = 88$ mm, and thus $h > hT$, and buckling controls the design. From (6.3), the shear buckling strength is found to be 29.1 MPa, giving a safety factor of 3.8.

To find whether interface delamination is of concern, the shear strain is substituted into (6.7), and the interfacial shear stress is found to be 3.7 MPa. Comparing with the nominal interfacial strength of 8.3 MPa from Table 6.6, it gives a safety factor of 2.3. The procedures can be repeated for other core heights, as shown in Table 6.11. It is interesting to note that several other panel heights, as shown in Table 6.11 in shaded areas, will fail due to inter-face delamination.

6.2.1.3 Facesheet Check

The bending moment is carried through the membrane forces of the facesheet, as shown in Figure 6.4. From Figure 6.8, we have

$$M = 2,768 \text{ kN-mm} \tag{6.19}$$

And the compressive and tensile forces (Figure 6.4) can be calculated as

$$C = T = \frac{M}{h} = \frac{2,768}{254} = 10.9 \text{ kN} \tag{6.20}$$

From experimental results, the compressive strength for the facesheet with a current configuration is 118.6 kN for $b = 25.4$ mm, giving a safety factor of 10.7. Safety factors for other laminate lay-ups (Table 6.1) are given in Table 6.12.

In conclusion, the example panel configuration is sufficient to sustain the given patch load.

6.2.2 Bridge System

6.2.2.1 Step 1: Equivalent Stiffness Properties of the FRP Deck

The equivalent properties of an FRP panel are listed in Table 6.13 based on procedures to obtain the equivalent engineering properties in Chapter 2 and Davalos et al. (2000).

TABLE 6.11

Shear Strength Check

Height (mm)	Aspect Ratio R	Shear Stiffness (MPa)	Shear Strain	Shear Stress in Flat Panel (MPa)	Buckling Strength (MPa)	Material Strength (MPa)	Controlling Strength (MPa)	Safety Factor for Flat Panel	Interfacial Normal Stress (MPa)	Safety Factor for Curved Panel
13	0.125	320.0	0.02189	92.1	1006.9	57.4	57.4	0.8	24.9	0.3
51	0.5	318.1	0.01101	46.3	137.8	57.4	57.4	1.5	18.1	0.5
102	1	316.5	0.00553	23.3	49.5	57.4	49.5	2.6	9.2	0.9
203	2	315.3	0.00278	11.7	31.0	57.4	31.0	3.3	4.6	1.8
254	2.5	315.0	0.00222	9.4	29.1	57.4	29.1	3.8	3.7	2.3
305	3	314.9	0.00185	7.8	28.4	57.4	28.4	4.5	3.1	2.7
508	5	314.7	0.00111	4.7	27.9	57.4	27.9	7.3	1.8	4.5
1016	10	314.5	0.00056	2.3	27.9	57.4	27.9	14.7	0.9	9.0

TABLE 6.12

Facesheet Check (see Table 6.1 for Laminate Descriptions)

	Laminate 1L	Laminate 1T	Laminate 2	Laminate 3
Failure load (kN)	118.6	62.0	104.0	97.4
Safety factor	10.7	5.7	9.7	9.0

TABLE 6.13

Equivalent Properties of FRP Panel

	E_x (GPa)	E_y (GPa)	v_x	G_{xy} (GPa)
In-plane	2.747	1.475	0.321	0.741
Bending	6.417	3.896	0.32	1.422

6.2.2.2 *Step 2: Effective Flange Width*

For interior girders, the effective flange width for a concrete deck based on AASHTO specification (2007) is the smaller value of:

1. One-fourth the effective span length: $0.25 \times 21{,}330 = 5{,}333$ mm
2. 12 times the average thickness of the slab, plus the greater of the web thickness or half the width of the top flange of the girder: $12 \times 254 \pm 201 = 3{,}249$ mm
3. The average spacing of adjacent beams: 2,440 mm

Therefore, $b_{eff\ AASHTO} = 2{,}440$ mm. Substituting this value and a DCA of 25% into (6.10) and (6.11), we have

$$b_{eff} = b_{eff\ AASHTO} \times R = 2440 \times 0.62 = 1513 \text{ mm} \tag{6.21}$$

6.2.2.3 *Step 3: Section Properties*

Based on b_{eff} calculated from (6.21), the total compression force that can be provided by the FRP deck can be calculated as

$$F_{deck-full} = b_{eff} \times 2 \times f_d = 1513 \times 2 \times 2.33 = 7051 \text{ kN} \tag{6.22}$$

where f_d is the compressive strength of the facesheet. It is assumed that the compressive force is carried by the two facesheets only. However, based on the spacing of the connector, there are a total of 18 connectors along the beam. The shear resistance for each connector is

$$Q_r = \varphi_{sc} \times Q_n = 0.85 \times 102 = 86.7 \text{ kN} \tag{6.23}$$

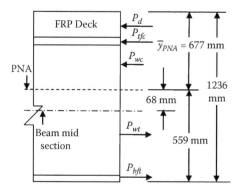

(a) Plastic forces for composite section

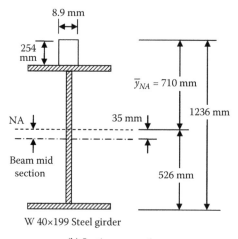

W 40×199 Steel girder

(b) Section properties

FIGURE 6.9
Partial composite sections.

where Q_n is the nominal capacity of the shear connector as described above and φ_{sr} is the resistance factor. The total shear force that can be transferred through the connectors is

$$F_d = 9 \times Q_r = 9 \times 86.7 = 780.3 \text{ kN} \tag{6.24}$$

The smaller value of $F_{deck\text{-}full}$ and F_d will be used in the design.

The plastic moment considering contribution from the FRP deck and steel girder can be calculated based on dimensions shown in Figure 6.9(a) and Table 6.14. By inspection, the plastic neutral axis (PNA) lies in the web of the steel beam, which is located a distance from the top of the deck; can be calculated to be 677 mm. Therefore;

TABLE 6.14

Dimensions of W40 × 199

Designation	Area (mm²)	Depth d (mm)	Web Thickness t_w (mm)	Flange		Moment of Inertia I (mm⁴)	Plastic Modulus Z_x (mm³)
				Width b_f (mm)	Thickness t_f (mm)		
W 40 × 199	37,677	982	17	400	27	6.202E+9	1.422E+07

- Bottom flange in tension:

  Force : $P_{bft} = F_y b_f t_{bf} = (345)(400)(27) = 3731$ kN   (6.25)

  Moment arm: $d_{bft} = 546$ mm

- Top flange in compression:

  Force : $P_{tfc} = F_y b_f t_{tf} = (345)(400)(27) = 3731$ kN   (6.26)

  Moment arm: $d_{tfc} = 409$ mm

- Tension web:

  Force : $P_{wt} = F_y b_{wt} t_w = (345)(532)(16.5) = 3032$ kN   (6.27)

  Moment arm: $d_{wt} = 266$ mm

- Compression web:

  Force : $P_{wc} = F_y b_{wc} t_w = (345)(396)(16.5) = 2251$ kN   (6.28)

  Moment arm: $d_{wc} = 198$ mm

- FRP deck:

  Force : $P_d = 780.3$ kN   (6.29)

  Moment arm: $d_d = 550$ mm

The plastic moment is the sum of the moments of the plastic forces about the PNA.

$$M_P = P_{bft}d_{bft} + P_{tfc}d_{tfc} + P_{wt}d_{wt} + P_{wc}d_{wc} + P_d d_d = 5.245 \times 10^6 \text{ kN-mm} \quad (6.30)$$

The modular ratio of n is calculated as

$$n = \frac{E_{steel}}{(E_y)_{deck}} = \frac{200\,GPa}{1.475\,GPa} = 136 \tag{6.31}$$

Therefore, the transformed width of the FRP deck is

$$b_{transformed} = \frac{b_{eff}}{n} = \frac{1513}{136} = 11.2 \text{ mm} \tag{6.32}$$

The transformed area is

$$A_{transformed} = b_{tranformed} \times 254 = 2845 \text{ mm}^2 \tag{6.33}$$

Based on the shear force transferred through the shear connector, the corresponding transformed area of the FRP deck is

$$A_{deck} = \frac{F_d}{F_y} = \frac{780.3 \times 10^3}{345} = 2262 \text{ mm}^2 \tag{6.34}$$

Since A_{deck} is less than $A_{tranformed}$, A_{deck} will be used to calculate the section properties, as shown in Table 6.15, based on dimensions shown in Table 6.14 and Figure 6.9(b).

The centroid of the section is calculated from the top of the FRP deck as

$$\bar{y}_{NA} = \frac{2.836 \times 10^7}{39941} = 710 \text{ mm} \tag{6.35}$$

The parallel axis theorem is used to get the moment of inertia of the components about this centroid, as shown in Table 6.15.

TABLE 6.15

Partial Composite Section Properties ($n = 136$, DCA = 25%)

Component	Width (mm)	Area (mm²)	y (mm)	Ay (mm³)	y-NA (mm)	A(y-NA)² (mm³)	I_0 (mm⁴)	I_x (mm⁴)
FRP deck	8.9	2263	127	2.874E+05	–583	3.029E+07	1.217E+07	7.816E+08
Steel		37677	745	2.807E+07	35	1.820E+06	6.202E+09	6.248E+09
Sum		39941		2.836E+07				7.030E+09

6.2.2.4 Step 4: Load Combinations

Strength I limit state:

$$U = \varphi[1.25DC + 1.50DW + 1.75(LL + IM)]$$

Service II limit state:

$$U = 1.0(DC + DW) + 1.0(LL + IM)$$

Fatigue and fracture limit state:

$$U = 0.75(LL + IM)$$

where DC = dead load of FRP deck and girders, DW = dead load of wearing surface, and $LL + IM$ = live load with dynamic allowance (impact).

6.2.2.5 Step 5: Live Load Effect

For this bridge with 9.75 m horizontal clearance, the number of lanes is selected as 2. The multiple presence factor is 1.0.

6.2.2.5.1 Strength Limit State

The dynamic allowance is 33%. The distribution factor for the moment can be calculated as 0.66 for interior girder with two lanes loaded following AASHTO specifications.

As shown in Figure 6.10, live load moments for truck, tandem, and lane load can be calculated as

$$M_{tr} = 145(5.33) + (145 + 35)(3.2) = 1328.7 \, kN - m \, (980 kips - ft)$$

$$M_{ta} = 110(5.33 + 4.72) = 1118.5 kN - m \, (825 kips - ft)$$

$$M_{ln} = \frac{9.3(21.3)^2}{8} = 531.5 kN - m \, (392 kips - ft)$$

Therefore, a design moment for the interior girder is

$$M_{LL+IM} = 0.66[1328.7(1.33) + 531.5] = 1517.2 kN - m \, (1119.0 kips - ft)$$

6.2.2.5.2 Service Limit State

The distribution factor for live load deflection is $DF = m(N_L/N_b) = 1*(2/5) = 0.4$.

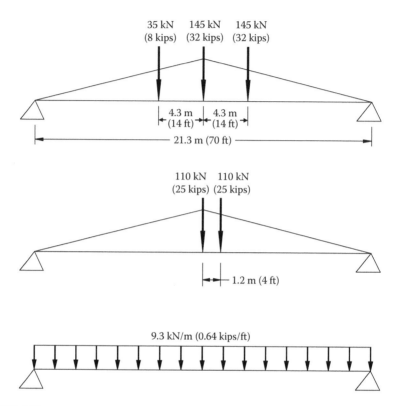

FIGURE 6.10
Truck, tandem, and lane load placement for maximum moment.

6.2.2.6 Step 6: Dead Load Effect

The dead load of a steel girder is 2.90 kN/m (0.199 kip/ft). The bridge FRP deck is assumed as 0.718 kN/m² (15 psf) with a wearing surface of 1.17 kN/m² (25 psf).

DC: Slab = 0.718*2.44 = 1.75 kN/m (0.12 kips/ft)
Steel girder = 2.90 kN/m
DW: Wearing surface = 1.17*2.44 = 2.92 kN/m (0.2kip/ft)

The induced maximum moments will be

$$M_{DC} = \frac{4.65(21.3)^2}{8} = 265.7 kN - m\ (196.0 kips - ft)$$

$$M_{DW} = \frac{2.92(21.3)^2}{8} = 166.1 kN - m\ (122.5 kips - ft)$$

6.2.2.7 Step 7: Strength Limit State Check for Interior Girder

Factored strength I moment:

$$M_u = 1.0[1.25(265.7) + 1.50(166.1) + 1.75(1517.2)] = 3236.4 kN-m \ (2387 kips-ft)$$

This rolled beam satisfies the compact section requirements. The plastic moment is

$$M_n = 5245 kN-m > M_u$$

6.2.2.8 Step 8: Service Limit State Check for Interior Girder

6.2.2.8.1 Live Load Deflection Control

1. From design truck alone:

   The front wheel load is $P_1 = 0.4*35*(1 + 0.33) = 19.12$ kN (4.3 kips).

   Each of the rear wheel loads is $P_2 = P_3 = 0.4*145*(1 + 0.33) = 72.62$ kN (17.0 kips).

   The corresponding deflection can be obtained from the AISC manual (2005).

   The live load deflection is $\Delta_{tr} = 22$ mm (0.86 in) $< \dfrac{span}{800} = 27$ mm.

2. For 25% of design truck the design lane load is:

   The live load deflection due to 25% truckload is $\Delta_{25\%tr} = 5$ mm (0.21 in).

   The live load deflection due to lane load is $\Delta_{ln} = 7$ mm (0.28 in).

   The total deflection is $\Delta_{total} = \Delta_{25\%tr} + \Delta_{ln} = 12$mm (0.45in) $< \dfrac{span}{800} = 27$mm.

6.2.2.8.2 Permanent Deflection Controls

Since no provision is provided for partial composite action for a permanent deflection check, noncomposite sections are assumed to be conservative. For both flanges of noncomposite sections

$$f_f = 0.80 R_h f_{yf} < 275.8 MPa$$

The service II moment is

$$M_s = 1.0(265.7 + 166.1) + 1.3(1517.2) = 2404.2 kN-m \ (1773.2 kips-ft)$$

$$f_f = \frac{M_s}{S} = \frac{2404.2 \times 10^6}{1.25 \times 10^7} = 192.3 MPa \ (27.6 ksi) < 275.8 MPa \ (40 ksi)$$

6.2.3 Discussions

Fatigue of the shear connector can be checked using the *S-N* curve shown in Figure 4.10, by assuming proper average daily truck traffic (ADTT). As illustrated from this example, an FRP deck with a low stiffness can still contribute to the stiffness and strength of the system. With 25% DCA and $n = 136$ ($n = 6$~10 for normal weight concrete), the stiffness and strength can be increased by 16 and 7%, respectively. Therefore, it is advantageous to include an FRP deck in the design by properly considering its DCA.

6.3 Conclusions

Design guidelines are provided for an HFRP sandwich panel with sinusoidal core geometry, and an example is given to illustrate the application of guidelines. It is expected that this study will not only contribute to the development of design specifications and facilitate the acceptance of this innovative lightweight structure, but also provide a base for developing design guidelines for other types of FRP sandwich structures.

References

AASHTO. (2007). *Load resistance and factor design, bridge design specifications*. 4th ed. American Association of State Highway and Transportation Officials, Washington, DC.

American Institute of Steel Construction. (2005). *Steel construction manual*. 13th ed. AISC, Chicago, IL.

Chen, A. (2004). Strength evaluation of honeycomb sandwich FRP panels with sinusoidal core geometry. PhD dissertation, Department of Civil and Environmental Engineering, West Virginia University, Morgantown.

Davalos, J.F., Qiao, P., Xu, X.F., Robinson, J., and Barth, K.E. (2000). Modeling and characterization of fiber-reinforced plastic honeycomb sandwich panels for highway bridge applications. *Composite Structures*, 52, 441–452.

7

Systematic Analysis and Design Approach for Single-Span FRP Deck-Stringer Bridges

7.1 Introduction

Previous chapters deal with fiber-reinforced polymer (FRP) deck–steel girder bridge systems. FRP decks can also be supported by FRP stringers. Therefore, in this chapter, a systematic approach for analysis and design of all FRP deck-stringer bridges (Qiao et al. 2000) is presented. This approach (Figure 7.1) is based on analyses at the microlevel (material), macrolevel (structural component), and system level (structure) to design all FRP deck-stringer bridge systems. First, based on manufacturer's information and material lay-up, ply properties are predicted by micromechanics. Once the ply stiffness properties are obtained, macromechanics is applied to compute the panel mechanical properties. Beam or stringer stiffness properties are then evaluated from mechanics of thin-walled laminated beams (MLB). Using elastic equivalence, apparent stiffness properties for composite cellular decks are formulated in terms of panel and single-cell beam stiffness properties, and their equivalent orthotropic material properties are further obtained. For design analysis of FRP deck-stringer bridge systems, an approximate series solution for the first-order shear deformation orthotropic plate theory is applied to develop simplified design equations, which account for load distribution factors for various load cases. As illustrated in Figure 7.1, the present systemic approach, which accounts for the microstructure of composite materials and geometric orthotropy of a deck system, can be used to design and optimize efficient FRP deck and deck-stringer systems.

To introduce this systemic approach shown in Figure 7.1, this chapter is organized in the following three main sections: (1) panel and beam analyses by micro/macromechanics and mechanics of thin-walled laminated beams, (2) FRP cellular decks by elastic equivalence analysis, and (3) analysis of a deck-stringer system by an approximate series solution technique. To verify the accuracy of the equivalent orthotropic material properties, a multi-box-beam deck fabricated by bonding side-by-side box FRP beams is experimentally tested and analyzed by a finite element model. To validate the approximate

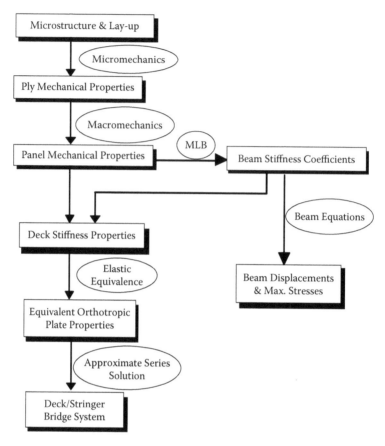

FIGURE 7.1
Systematic analysis protocol for FRP bridge systems.

series solution, the multi-box-beam deck is attached to FRP wide-flange (WF) beams and tested and analyzed as a deck-stringer bridge system. The box beams for decks and WF beams for stringers were both produced by pultrusion. Both deck and bridge systems are tested under static loads for various load conditions. To illustrate the systematic design procedures developed in this chapter, an example of an FRP deck-stringer bridge is presented.

7.2 Panel and Beam Analysis

Extensive research has been conducted in the area of analysis and design of composite materials at micro- and macrolevels. The analysis of FRP beams from micro/macromechanics to beam response has been presented in Davalos et al. (1996a). In this section, the analyses of micro/macrostructure and beam component are briefly reviewed and include (1) constituent materials and prediction of ply properties, (2) laminated panel engineering properties, and (3) beam or stringer stiffness properties.

7.2.1 Panel Analysis by Micro/Macromechanics

Although pultruded FRP shapes are not laminated structures in a rigorous sense, they are pultruded with material architectures that can be simulated as laminated configurations. A typical pultruded section mainly includes the following three types of layer (Davalos et al. 1996a) (see Figures 7.2 and 7.3): (1) continuous strand mats (CSMs), (2) stitched fabrics (SFs), and (3) rovings or unidirectional fiber bundles. Each layer is modeled as a homogeneous, linearly elastic, and generally orthotropic material. Based on information provided by the manufacturer, the fiber volume fraction (V_f) can be evaluated and used to compute the ply stiffness properties from micromechanics models (Luciano and Barbero 1994). For the box section of Figure 7.2 and wide-flange section of Figure 7.3, the predicted ply properties are given in Tables 7.1 and 7.2, respectively, computed using the micromechanic model from Luciano and Barbero (1994). Once the ply stiffness

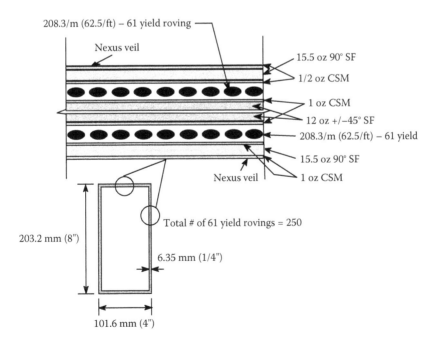

FIGURE 7.2
Microstructure and dimensions of FRP box beam section.

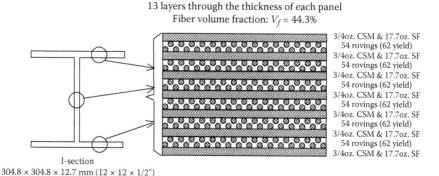

13 layers through the thickness of each panel
Fiber volume fraction: V_f = 44.3%

3/4oz. CSM & 17.7oz. SF
54 rovings (62 yield)
3/4oz. CSM & 17.7oz. SF
54 rovings (62 yield)
3/4oz. CSM & 17.7oz. SF
54 rovings (62 yield)
3/4oz. CSM & 17.7oz. SF
54 rovings (62 yield)
3/4oz. CSM & 17.7oz. SF
54 rovings (62 yield)
3/4oz. CSM & 17.7oz. SF
54 rovings (62 yield)
3/4oz. CSM & 17.7oz. SF

I-section
304.8 × 304.8 × 12.7 mm (12 × 12 × 1/2")

FIGURE 7.3
Panel fiber architectures of wide-flange beam.

TABLE 7.1

Ply Material Properties of Box Section (Figure 7.2)

Lamina	E_1 (× 10^4 MPa)	E_2 (× 10^4 MPa)	v_{12}	G_{12} (× 10^4 MPa)
½ oz CSM	1.4	1.4	0.407	0.5
1 oz CSM	1.2	1.2	0.402	0.4
15.5 oz 90° SF	2.8	0.8	0.389	0.3
12 oz ± 45° SF	2.4	0.7	0.396	0.3
61 yield roving	5.9	2.3	0.343	1.0

TABLE 7.2

Material Properties of Wide-Flange Section (Figure 7.3)

Lamina	E_1 (× 10^4 MPa)	E_2 (× 10^4 MPa)	v_{12}	G_{12} (× 10^4 MPa)
¾ oz CSM	1.2	1.2	0.402	0.4
17.7 oz ± 45° SF	2.9	0.8	0.294	0.3
62 yield roving	4.7	1.4	0.278	0.6

properties for each laminate (panel) of an FRP beam are computed, the stiffness properties of a laminated panel can be computed from macrome-chanics (Barbero 1999). For example, for the box beam shown in Figure 7.2, the panel properties (E_x, E_y, v_{xy}, and G_{xy}) predicted by the micro/macrome-chanics models (Luciano and Barbero 1994) correlate well with experimen-tal results for coupon samples (Table 7.3) tested in tension and torsion.

TABLE 7.3

Panel Properties of Box Section (10.16 × 20.32 × 0.635 cm (4 × 8 × ¼ in.))

	E_x	E_y	v_{xy}	G_{xy}
Experimental	2.3×10^4 MPa[a]	1.7×10^4 MPa[a]	0.269[a]	59.5×10^4 MPa
Micro/macromechanics	2.3×10^4 MPa	1.8×10^4 MPa	0.285	60.6×10^4 MPa
% difference	+2.6%	+5.2%	+5.9%	+1.9%

[a] From tension tests.
[b] From torsion tests.

7.2.2 Beam Analysis by Mechanics of Laminated Beams

The response of FRP shapes in bending is evaluated using the mechanics of thin-walled laminated beams (MLB) (Barbero et al. 1993). In MLB, the stiffness coefficients (axial, A; bending, D; axial-bending coupling, B; and shear, F) of a beam are computed by adding the contributions of the stiffness properties of the component panels, which in turn are obtained from the effective beam moduli. Based on MLB, engineering design equations for FRP beams under bending have been formulated (Davalos et al. 1998), and they can be easily adopted by practicing engineers and composite manufacturers for the analysis, design, and optimization of structural FRP beams or bridge stringers. MLB is suitable for straight FRP beams or columns with at least one axis of geometric symmetry and can be used to evaluate the stiffness properties and response of bridge stringers. As an example, the bending (D) and shear (F) stiffnesses of box beam (Figure 7.2) and wide-flange beam (Figure 7.3) by MLB are listed in Table 7.4, and experimental results for deflections and strains compared favorably with MLB predictions (Salim et al. 1995a; Davalos et al. 1996a, 1998).

The panel and beam properties obtained above by micro/macromechanics and MLB can be efficiently implemented in deck and deck-stringer system design, as described in Sections 7.3 and 7.4.

TABLE 7.4

Strong-Axis Beam Bending and Shear Stiffness Coefficients by MLB

Beam Stiffness	D_b (N/m²-m⁴)	F_b (N/m²-m²)
Box (10.16 × 20.32 × 0.635 cm (4 × 8 × ¼ in.))	5.16E+05	1.55E+07
WF (30.48 × 30.48 × 1.27 cm (12 × 12 × ½ in.))	4.91E+06	2.24E+07

7.3 FRP Cellular Decks: Elastic Equivalence

A multicellular FRP composite bridge deck can be modeled as an orthotropic plate, with equivalent stiffnesses that account for the size, shape, and constituent materials of the cellular deck. Thus, the complexity of material anisotropy of the panels and structural orthotropy of the deck system can be reduced to an equivalent orthotropic plate with global elastic properties in two orthogonal directions: parallel and transverse to the longitudinal axis of the deck cell. These equivalent orthotropic plate properties can be directly used in design and analysis of a deck-stringer bridge system, as presented in Section 7.4, and they can also serve to simplify modeling procedures in either numerical or explicit formulations. The design equations necessary for such a model are presented in this section, along with numerical and experimental verification of the results.

In this section, the development of equivalent stiffnesses for cellular decks consisting of multiple FRP box beams is presented. Multicell box sections are commonly used in deck construction because of their light weight, efficient geometry, and inherent stiffness in flexure and torsion. Also, this type of deck has the advantage of being relatively easy to build. It can be either assembled from individual box beams or manufactured as a complete section by pultrusion or vacuum-assisted resin transfer molding process. The elastic equivalence approach (Troitsky 1987) used in this chapter accounts for out-of-plane shear effects, and the results for a multicell box section are verified experimentally and by finite element analyses.

7.3.1 Equivalent Stiffness for Cellular FRP Decks

As an illustrative example, we derive the bending, shear, and torsional equivalent stiffnesses for a deck composed of multiple box sections (Figure 7.4).

7.3.1.1 Longitudinal Stiffnesses of Cellular FRP Deck

The bending stiffness of the deck in the longitudinal direction, or x axis in Figure 7.4, is expressed as the sum of the bending stiffness of individual box beams (Db; see Table 7.4):

$$D_x = n_c D_b \tag{7.1}$$

where n_c = number of cells. For the section shown in Figure 7.4, b = width of a cell, h = height of a cell, t_f = thickness of the flange, and t_w = thickness of the web. If all panels have identical material lay-up and $t_f = tw = t$, (7.1) becomes

$$D_x = n_c E_x \left(h^2 + t^2 + 3(h)b \right) \frac{(h)(t)}{6} \tag{7.2}$$

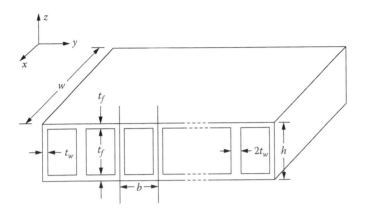

FIGURE 7.4
Geometric parameters of multicell box deck.

where E_x = modulus of elasticity of a panel in the x direction computed by micro/macromechanics or obtained experimentally (Table 7.3).

The out-of-plane shear stiffness of the deck in the longitudinal direction, F_x, is expressed as a function of the stiffness for the individual beams (F_b):

$$F_x = n_c F_b \tag{7.3}$$

where F_b is given in Table 7.4, and n_c = number of cells. This expression can be further approximated in terms of the in-plane shear modulus of the panel, G_{xy} (see Table 7.3), and cross-sectional area of the beam webs:

$$F_x = n_c G_{xy}(2t)h \tag{7.4}$$

7.3.1.2 Transverse Stiffnesses of Cellular FRP Deck

An approximate value for the deck bending stiffness in the transverse direction, D_y, may be obtained by neglecting the effect of the transverse diaphragms and the second moment of area of the flanges about their own centroids. For a deck as shown in Figure 7.4, with $t_f = t$:

$$D_y = \frac{1}{2} E_y(w)(t)h^2 \tag{7.5}$$

where w is the length of the deck in the longitudinal direction and E_y is the modulus of elasticity of the panel in the y direction (Table 7.3).

For multiple box sections, the simplest way to obtain the deck's out-of-plane transverse shear stiffness is to treat the structure as a Vierendeel frame in the transverse direction (Cusen and Pama 1975). For the Vierendeel frame (Figure 7.5), the inflection points are assumed at the midway of top

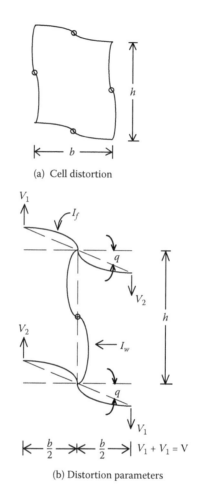

(a) Cell distortion

(b) Distortion parameters

FIGURE 7.5
Vierendeel distortion in multicell box beam.

and bottom flanges between the webs. The shear stiffness in the transverse direction, F_y, for the cross section shown in Figure 7.4 may be written as

$$F_y = \frac{V}{\theta} = \frac{12 E_y}{b\left(\dfrac{h}{I_w} + \dfrac{b}{2 I_f}\right)} \qquad (7.6)$$

where the moments of inertia I are defined as

$$I_f = \frac{w t_f^3}{12}; \quad I_w = \frac{w(2 t_w)^3}{12} \qquad (7.7)$$

For $t_f = t_w = t$, (7.6) can be simplified as

$$F_y = \frac{2E_y w t^3}{b\left(b+\dfrac{h}{4}\right)}$$ (7.8)

where E_y is the modulus of elasticity of a panel in the y direction (Table 7.3).

7.3.1.2.1 Torsional Stiffness of Cellular FRP Deck

The torsional rigidity of a multicell section, GJ, is evaluated by considering the shear flow around the cross section of a multicell deck. For a structure where the webs and flanges are small compared with the overall dimensions of the section, Cusens and Pama (1975) have shown that the torsional rigidity may be written as

$$GJ = \frac{4A^2 G_{xy}}{\sum \dfrac{ds}{t}} + \sum G_{xy}(ds)\frac{t^3}{3}$$ (7.9)

where A = area of the deck section including the void area and is defined as $A = n_c b h$, and $\Sigma ds/t$ represents the summation of the length-to-thickness ratio taken around the median line of the outside contour of the deck cross section. For a constant panel thickness t, the torsional rigidity can be simplified as

$$GJ = \frac{2(n_c b h)^2 G_{xy} t}{(n_c b + h)} + \frac{2}{3}(n_c b + h)G_{xy} t^3$$ (7.10)

The above approximate equation is justified by the fact that for a multicell deck, the net shear flows through interior webs are negligible, and only the shear flows around the outer webs and top and bottom flanges are significant. The second term in (7.10) is relatively small compared to the first term and can be ignored.

If the deck is treated as an equivalent orthotropic plate, its torsional rigidities depend upon the twist in two orthogonal directions. Thus torsional stiffness D_{xy} may be taken as one-half of the total torsional rigidity given by (7.10) divided by the total width of the deck:

$$D_{xy} = \frac{GJ}{2n_c b}$$ (7.11)

Substituting (7.10) into (7.11) and neglecting the second term in (7.10), we get

$$D_{xy} = \frac{n_c G_{xy} b h^2 t}{(n_c b + h)}$$ (7.12)

where D_{xy} is the torsional stiffness per unit width (N-m^3/m^2 (lb-in.^{3}/in.^{2})).

7.3.2 Verification of Deck Stiffness Equations by Finite Element Analysis

The formulas for bending and torsional stiffnesses obtained in Section 7.3.1 are based on the assumption that the deck system behaves as a beam and does not account for the Poisson effects of the deck. To verify the accuracy of the above deck stiffness equations, a finite element analysis of the deck system is performed. The model is shown in Figure 7.4 and consists of box beams (Figure 7.2) bonded side by side to form an integral deck. The computer program NISA (1994) is used, and the panels are modeled with eight-node isoparametric layered shell elements. The cellular decks subject to line loading for longitudinally supported and transversely supported conditions are shown in Figures 7.6 and 7.7, and the model for torsional loading is given in Figure 7.8.

7.3.2.1 Verification of Bending and Shear Stiffnesses

The deck bending and shear stiffnesses in the longitudinal and transverse directions are used to evaluate mid-span deflections from the following:

$$\delta_3 = \frac{PL^3}{48D_i} + \frac{PL}{4\kappa F_i} \quad \text{(3-point bending)} \tag{7.13}$$

$$\delta_4 = \frac{23PL^3}{1296D_i} + \frac{PL}{6\kappa F_i} \quad \text{(4-point bending)} \tag{7.14}$$

where P = total applied load, L = span length, κ = shear correction factor ($\kappa \cong 1.0$ is assumed in the analysis), and D_i and F_i = bending and shear stiffness ($i = x$ for longitudinal or y for transverse directions). The deflections by (7.13) and (7.14) in terms of stiffness properties are compared with results from the finite element model for actual cellular systems under line loading (Figures 7.6 and 7.7). For the longitudinal stiffness verification, the length of the decks is kept constant ($L = 274.32$ cm (108 in.)), and the deflection in terms of bending and shear stiffnesses is a function of the number of cells. Each deck is simply supported and subjected to either three-point or four-point bending due to uniformly distributed line loads. The comparisons between the predictions of (7.13) and (7.14) based on simplified stiffness formulas and the finite element results for actual decks are presented graphically in Figure 7.9.

Similarly, the mid-span deflections in the transverse direction are found by modeling several multicellular decks comprised of $10.16 \times 20.32 \times 0.635$ cm ($4 \times 8 \times \frac{1}{4}$ in.) box sections (Figure 7.7). For these models, the width (w) is kept constant ($w = 30.48$ cm (12 in.)), and the deflection is a function of the number of cells. The model is simply supported and subjected to either three-point or four-point bending due to uniformly distributed line loads. The results of the finite element models and theoretical predictions are shown in Figure 7.10.

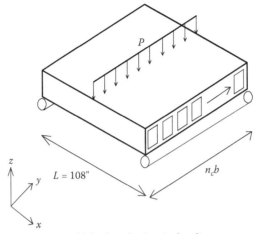

(a) Deck under 3-point bending

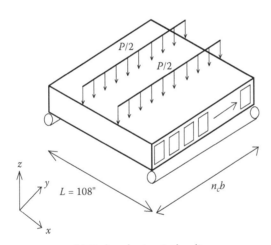

(b) Deck under 4-point bending

FIGURE 7.6
FE model for verification of longitudinal bending and shear stiffness equations.

7.3.2.2 Verification of Torsional Stiffness of the Deck

The simplified formula for the torsional rigidity, GJ, of the deck system was also verified using finite element analyses, which indirectly serve to verify the torsional stiffness of the deck (D_{xy}). The model shown in Figure 7.8 consisted of a multicellular deck with one end fixed, by constraining displacements and rotations in all three principal directions and all three rotations, and the other end subjected to a uniform torque. The longitudinal torsional rigidity of a deck is expressed in terms of the angle of twist ϕ and the torque applied at the end of the section as

$$GJ = \frac{TL}{\phi} \qquad (7.15)$$

where $T = 2qn_cbh$ (as shown in Figure 7.8) is the applied torque. The specimen length L is held constant ($L = 274.32$ cm (108 in.)), and the number of cells is used as the design variable. The finite element results are compared with the theoretical predictions of (7.10), and the results are presented in Figure 7.11.

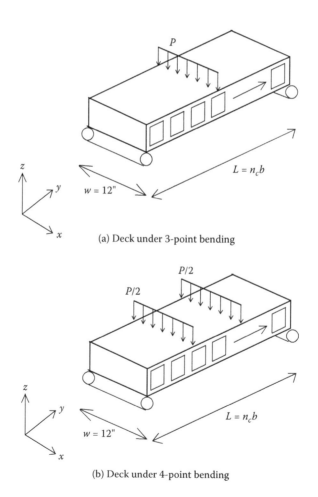

(a) Deck under 3-point bending

(b) Deck under 4-point bending

FIGURE 7.7
FE model for verification of transverse bending and shear stiffness equations.

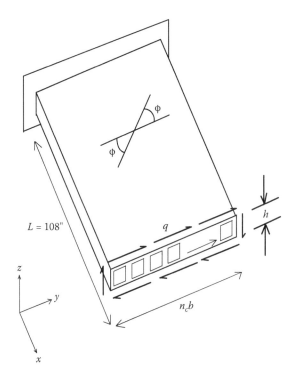

FIGURE 7.8
Model for verification of torsional rigidity equation.

7.3.2.3 Comparisons and Remarks

As shown in Figures 7.9 to 7.11, a good correlation is obtained between the theoretical predictions based on the simplified stiffness formulas and the finite element analyses of an actual deck. For the deflection in terms of longitudinal stiffnesses (D_x and F_x), the maximum percent difference is 4%, and for the deflection in terms of transverse stiffnesses (D_y and F_y), the maximum difference is about 10%. For the longitudinal torsional stiffness, the discrepancy of results increases steadily from 6% for one1 cell to 22% for 15 fifteen cells. Some limited experimental data available for one and two cells (Salim 1997) match closely the analytical results. The favorable deflection comparisons between beam equations and finite element results indirectly verify the accuracy of the deck bending stiffness equations. Similarly, the torsion results indicate that the simplified torsional stiffness equations are acceptable for practical applications. Therefore, the proposed relatively simple stiffness equations account for both shape and material anisotropy of the deck and can be used with relative confidence in design analysis of cellular bridge deck systems.

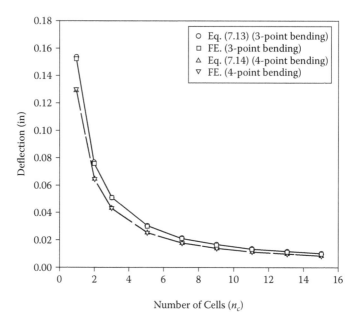

FIGURE 7.9
Longitudinal central deflection of a multicell deck (Figure 7.6).

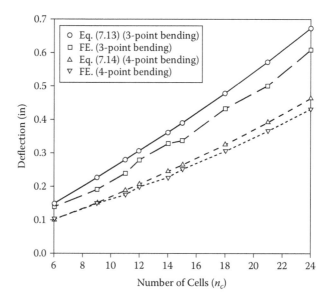

FIGURE 7.10
Transverse central deflection of a multicell deck (Figure 7.7).

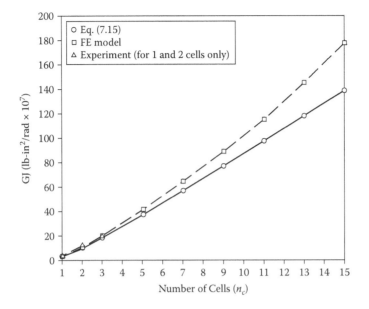

FIGURE 7.11
Torsional rigidity vs. number of cells (Figure 7.8).

7.3.3 Equivalent Orthotropic Material Properties

Once the stiffness properties of an actual deck are obtained, it is a simple matter to calculate effective material properties for an equivalent orthotropic plate. To obtain the equivalent orthotropic plate material properties for an actual deck, we can further simplify the design analysis of deck and deck-stringer bridge systems.

To calculate the moduli of elasticity $(Ex)_p$ and $(Ey)_p$ for the equivalent orthotropic plate, the relationship $D = EI$ is used, leading to

$$(E_x)_p = 12\frac{D_x}{t_p^3 b_p}(1-v_{xy}v_{yx}) \tag{7.16}$$

$$(E_y)_p = 12\frac{D_y}{t_p^3 l_p}(1-v_{xy}v_{yx}) \tag{7.17}$$

where the subscript p indicates property related to the equivalent orthotropic plate, t_p = thickness of the plate (= h for the actual deck, Figure 7.4), b_p = width of the plate (= $n_c b$ for the actual deck), and l_p = length of the plate (= w for the actual deck). The Poisson's ratios v_{ij} are defined as

$$v_{ij} = -\frac{\varepsilon_j}{\varepsilon_i} \tag{7.18}$$

where ε is the strain in the i or j direction. For orthotropic materials, the Poisson's ratio must obey the following relationship:

$$\frac{v_{ij}}{E_i} = \frac{v_{ji}}{E_j} \quad \text{or} \quad \frac{v_{xy}}{D_x} = \frac{v_{yx}}{D_y} \tag{7.19}$$

In this study, we use the approximation $v_{xy} = 0.3$, which is typically used for pultruded composites.

To calculate the out-of-plane shear moduli $(G_{xz})_p$ and $(G_{yz})_p$, the relationship $F = GA$ is used, leading to

$$(G_{xz})_p = \frac{F_x}{t_p b_p} \tag{7.20}$$

$$(G_{yz})_p = \frac{F_y}{t_p l_p} \tag{7.21}$$

Finally, to calculate the in-plane shear modulus $(G_{xy})_p$, we use

$$(G_{xy})_p = 6\frac{D_{xy}}{t_p^3} \tag{7.22}$$

With these equivalent material properties, it is now easy to use explicit plate solutions (see Section 7.4) for analysis and design of cellular decks.

7.3.4 Experimental and Numerical Verification of Equivalent Orthotropic Material Properties

To indirectly verify the accuracy of the equivalent orthotropic material properties given in (7.16) through (7.22) (see Section 7.3.3), a multi-box-beam deck of 1.524 m × 2.74 m × 20.32 cm (5 ft × 9 ft × 8 in.) (Figure 7.12a) subjected to a patch load is tested and analyzed for three load conditions: (1) at the center of the deck, (2) at 16 in. to one side from the center along the line AA′, and (3) at 16 in. to the other side from the center along the line AA′. The finite element program NISA (1994) is used to conduct two distinct analyses: (1) the actual deck (Figure 7.13) is modeled using eight-node isoparametric layered shell elements and the material properties of Table 7.1; (2) an equivalent solid orthotropic plate of the same global dimensions as the actual deck is modeled using the same elements and the equivalent material properties

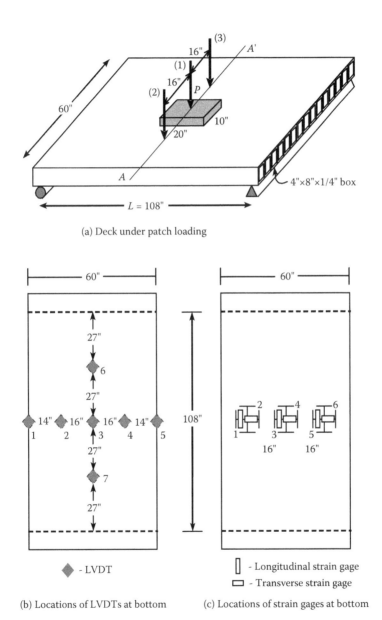

(a) Deck under patch loading

(b) Locations of LVDTs at bottom

(c) Locations of strain gages at bottom

FIGURE 7.12
Experimental setup of multicell box deck (1.52 m ´ 3.05 m ´ 20.32 cm (5 ft ´ 10 ft ´ 8 in.)).

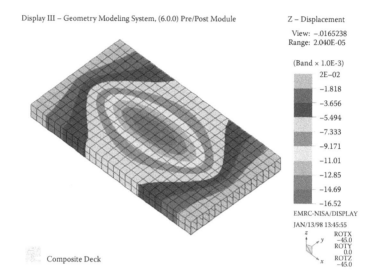

Composite Deck

FIGURE 7.13
FE simulation and deflection contour of multicell box deck.

computed from (7.16) through (7.22) and given in Table 7.5. The experimental results and correlations with finite element analyses are presented next.

7.3.4.1 Experimental Details

The test sample was fabricated by bonding FRP box beams side by side with epoxy (Brown 1998). For each load condition, the displacements are recorded at several locations with linear variable displacement transducers (LVDTs) (see Figure 7.12), and the strains in the longitudinal and transverse directions are obtained at three locations by bonding 350 Ω strain gages at the bottom of the deck (Figure 7.12(c)). Note from Figure 7.12 that for the asymmetric load cases 2 and 3, the following displacement values should be approximately equal: δ_1 and δ_5, and δ_2 and δ_4. Similarly, the following strains should correspond to each other: ε_1 and ε_5, and ε_2

TABLE 7.5

Deck Stiffness Properties and Orthotropic Material Properties for Cellular Deck 1.5 m × 2.7 m × 20.3 cm (5 ft × 9 ft × 8 in.)

D_x (N-m⁴/m²)	D_y (N-m⁴/m²)	v_{xy}	D_{xy} (N-m⁴/m²)	F_x (N-m²/m²)	F_y (N-m²/m²)
7.74E+06	6.47E+06	0.3	3.32E+04	2.18E+08	1.63E+06
E_x (MPa)	E_y (MPa)	v_{yx}	G_{xz} (MPa)	G_{xz} (MPa)	G_{yz} (MPa)
6.72E+03	3.12E+03	0.25	9.34E+02	4.65E+01	2.16E+01

7.3.4.2 Experimental Results and Correlation

Comparisons of experimental results and finite element analyses for FRP decks under centric loading (load case 1) and asymmetric loading (load cases 2 and 3) are shown in Tables 7.6 and 7.7, respectively. Tables 7.6 and 7.7 indicate that the measured displacements and strains compare relatively well with

TABLE 7.6

Experimental and Finite Element Comparison for Multicell Box Deck under Load Case 1 (Centric)

Parameter	Experiment	FE (Actual Deck)	FE Equivalent Plate
δ_1 (mm/kN)	0.04107	0.03572	0.03429
δ_2 (mm/kN)	0.05531	0.04904	0.05394
δ_3 (mm/kN)	0.11598	0.09365	0.10823
δ_4 (mm/kN)	0.05491	0.04904	0.05394
δ_5 (mm/kN)	0.04044	0.03560	0.03435
δ_6 (mm/kN)	0.08795	0.05884	0.07075
δ_7 (mm/kN)	0.08094	0.05884	0.07075
ε_1 ($\mu\varepsilon$/kN)	6.30420	6.38314	7.12073
ε_2 ($\mu\varepsilon$/kN)	−3.29686	−1.30789	−1.75215
ε_3 ($\mu\varepsilon$/kN)	15.38071	13.68956	16.49193
ε_4 ($\mu\varepsilon$/kN)	−3.42357	−5.9478	−5.30220
ε_5 ($\mu\varepsilon$/kN)	6.36027	6.38314	7.12073
ε_6 ($\mu\varepsilon$/kN)	−2.90777	−1.30789	−1.75215

TABLE 7.7

Finite Element Comparison for Multi-Cell Box Deck under Load Cases 2 and 3 (Asymmetric)

Load Cases 2 and 3	Load Case 2	Load Case 3	Load Cases 2 and 3	Actual Deck	Equivalent Plate
δ_1 and δ_5 (mm/kN)	−0.0109	−0.3082	−0.1596	0.0054	−0.0067
δ_2 and δ_4 (mm/kN)	0.0170	0.0235	0.0203	0.0201	0.0144
δ_3 and δ_3 (mm/kN)	0.0510	0.0621	0.0566	0.0490	0.0513
δ_4 and δ_2 (mm/kN)	0.1218	0.1327	0.1273	0.1105	0.1284
δ_5 and δ_1 (mm/kN)	0.1062	0.1083	0.1073	0.0985	0.1198
δ_6 and δ_6 (mm/kN)	0.0455	0.0567	0.0227	0.0347	0.0360
δ_7 and δ_7 (mm/kN)	0.0431	0.0497	0.0464	0.0322	0.0360
ε_1 and ε_5 ($\mu\varepsilon$/kN)	5.5285	5.3262	5.4273	2.8764	1.8674
ε_2 and ε_6 ($\mu\varepsilon$/kN)	−3.7259	−5.2246	−4.4754	−0.4880	−0.1552
ε_3 and ε_3 ($\mu\varepsilon$/kN)	6.2497	6.5684	6.4092	6.3977	6.7301
ε_4 and ε_4 ($\mu\varepsilon$/kN)	−4.8333	−6.9983	−5.9158	−1.1406	−1.3774
ε_5 and ε_1 ($\mu\varepsilon$/kN)	17.9507	18.6780	18.3145	15.7153	19.0180
ε_6 and ε_2 ($\mu\varepsilon$/kN)	−6.0712	−7.7296	−6.9005	−6.6031	−6.0084

FE models of actual deck and equivalent plate for both the symmetric loading (case 1) and asymmetric loading (cases 2 and 3) cases. For the symmetric load case (case 1), the difference of deflection (δ_3) under load point between the experiment and the FE equivalent plate model is about 6.5%, whereas there is a 6.7% difference for longitudinal strain (ε_3) at the center of the deck (Figure 7.12). As noted in Table 7.7, values for the asymmetric loading (cases 2 and 3) also compare favorably between the average experiment data and FE equivalent plate model when the measurements are close to the applied load; the differences are about 0.9% for deflection and 5.0% for longitudinal strain under applied load. The good correlation between experiment and FE models validates the orthotropic plate material properties obtained by elastic equivalence analysis, which can be used next in analysis of an FRP deck-stringer system.

7.4 Analysis of FRP Deck-Stringer Bridge System

The equivalent properties for cellular decks and stiffnesses for FRP beams can be efficiently used to analyze and design deck-stringer systems. We present in this section an overview of a series solution for stiffened orthotropic plates based on first-order shear deformation theory and transverse interaction of forces between the deck and the stringers. The solutions for symmetric and antisymmetric load cases are used to obtain the solution for asymmetric loading. Based on deck-stringer transverse interaction force functions, wheel load distribution factors are derived, which are used later to provide design guidelines for deck-stringer bridge systems. Finally, the approximate series solution is verified by testing a 3.048 m × 3.048 m × 20.32 cm (10 ft × 10 ft × 8 in.) multi-box-beam deck supported by WF 30.48 × 30.48 × 1.27 cm (12 × 12 × ½ in. FRP) beams; this system is also analyzed by the finite element model (NISA 1994).

7.4.1 First-Order Shear Deformation Theory for FRP Composite Deck

A first-order shear deformation theory (Reddy 1984) is applied to analyze the behavior of a geometrically orthotropic FRP composite deck. Instead of direct modeling of the actual deck geometry, an equivalent orthotropic plate, as discussed in Section 7.3, is used to simplify the analysis. The formulas for equivalent orthotropic material properties accounting for deck geometry and panel-laminated material properties are given in Section 7.3. The equilibrium equations accounting for first-order shear deformation of an orthotropic plate are:

$$A_{55} \frac{\partial}{\partial x}\left(\psi_x + \frac{\partial w_o}{\partial x}\right) + A_{44} \frac{\partial}{\partial y}\left(\psi_y + \frac{\partial w_o}{\partial y}\right) + q(x,y) = 0$$

$$\frac{\partial}{\partial x}\left(D_{11} \frac{\partial \psi_x}{\partial x} + D_{12} \frac{\partial \psi_y}{\partial y}\right) + D_{66} \frac{\partial}{\partial y}\left(\frac{\partial \psi_x}{\partial y} + \frac{\partial \psi_y}{\partial x}\right) - A_{55}\left(\psi_x + \frac{\partial w_o}{\partial x}\right) = 0 \quad (7.23)$$

$$D_{66} \frac{\partial}{\partial x}\left(\frac{\partial \psi_x}{\partial y} + \frac{\partial \psi_y}{\partial x}\right) + \frac{\partial}{\partial y}\left(D_{12} \frac{\partial \psi_x}{\partial x} + D_{22} \frac{\partial \psi_y}{\partial y}\right) - A_{44}\left(\psi_y + \frac{\partial w_o}{\partial y}\right) = 0$$

where A_{ij} $(i,j = 4,5)$ are the intralaminar shear stiffnesses, and D_{ij} $(i,j = 1,2,6)$ are the bending stiffnesses for an orthotropic material.

The deck-stringer bridge system can first be analyzed as an orthotropic plate stiffened by edge stringers (or beams) (Salim et al. 1997). Then the contributions of interior stringers are accounted for in the formulation by considering the interaction forces and the comparability conditions along rib lines between the deck and stringers. The analysis is general with respect to (1) size and stiffness of the deck, and (2) type of loading (uniform or concentrated). The formulation is concerned first with symmetric and antisymmetric loading conditions.

7.4.1.1 System under Symmetric Loading Case

A Fourier polynomial series is employed to obtain the solutions for the equilibrium equations (7.23). The solution for a symmetric loading is

$$w_o(x,y) = \sum_{i,j=1}^{\infty} W_{ij} \sin \alpha x (\sin \beta y + W_o)$$

$$\psi_x(x,y) = \sum_{i,j=1}^{\infty} X_{ij} \cos \alpha x (\sin \beta y + X_o) \quad (7.24)$$

$$\psi_y(x,y) = \sum_{i,j=1}^{\infty} Y_{ij} \sin \alpha x \cos \beta y$$

where $\alpha = i\pi/a$ and $\beta = j\pi/b$, and W_{ij}, X_{ij}, and Y_{ij} are the coefficients to be determined to complete the solution. Note that these series approximations satisfy the essential boundary conditions. The generalized loading can be written as the following infinite double series:

$$q(x,y) = \sum_{i,j=1}^{\infty} Q_{ij} \sin \alpha x \sin \beta y \quad (7.25)$$

Q_{ij} are the Fourier coefficients in the representation of the load $q(x, y)$. By substituting the general solutions (7.24) and (7.25) into (7.23) and reducing by orthogonality conditions (Salim et al. 1995b; Salim 1997), we obtain the following system of equations for any number of terms (i, j):

$$
\begin{bmatrix}
K_{11} & K_{12} & K_{13} \\
K_{21} & K_{22} & K_{23} \\
K_{13} & K_{23} & K_{33}
\end{bmatrix}
\begin{Bmatrix}
W_{ij} \\
X_{ij} \\
Y_{ij}
\end{Bmatrix}
=
\begin{Bmatrix}
Q_{ij} \\
0 \\
0
\end{Bmatrix}
\tag{7.26}
$$

where K_{ij} are the deck stiffness coefficients (for a symmetric loading) (Brown 1998). For a one-term approximation, the constants W_0 and X_0 are obtained by satisfying the boundary conditions of the edge-stiffened orthotropic plate (Figure 7.14(b)):

$$
W_0 = A_{44} C \left(\frac{Y_{11}}{W_{11}} + \beta \right)
$$

$$
X_0 = -\frac{A_{44}}{\alpha^3 D} \left(\frac{Y_{11} + \beta W_{11}}{X_{11}} \right)
\tag{7.27}
$$

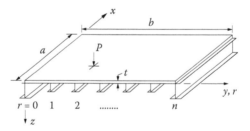

(a) Deck-and-stringer system

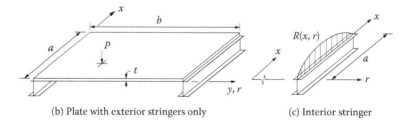

(b) Plate with exterior stringers only   (c) Interior stringer

FIGURE 7.14
Deck-stringer bridge system.

where

$$c = \frac{1}{\alpha^2}\left(\frac{1}{\kappa F} + \frac{1}{\alpha^2 D}\right),$$

κ = the stringer shear correction factor, and F and D are, respectively, the shear and bending stiffnesses of the stringer and are obtained based on mechanics of laminated beams (MLB) (Table 7.4) (Barbero et al. 1993). For any interior stringer at any location r ($r = 0, 1, \ldots, n$) (see Figure 7.14(c)), the generalized deflection function for any symmetric loading is (Brown 1998)

$$w^R(x,r) = R_{11}\frac{1}{\alpha^2}\left(\frac{1}{\kappa F} + \frac{1}{\alpha^2 D}\right)\sin\frac{\pi x}{a}\left(\sin\frac{\pi r}{n} + W_o\right) \tag{7.28}$$

where

$$R_{11} = \frac{Q_{11}}{\frac{1}{\alpha^2}\left(\frac{1}{\kappa F} + \frac{1}{\alpha^2 D}\right)\frac{Q_{11}}{W_{11}} + \frac{n}{b}\left(1 + \frac{4W_o}{\pi}\right)}$$

7.4.1.2 System under Antisymmetric Loading Case

Analogous to the symmetric case, (7.24) and (7.25) are modified for a first-term approximation of an antisymmetric loading as

$$w_o(x,y) = W_{12}\sin\alpha x\left(\sin 2\beta y + W_1\left(1 - \frac{2y}{b}\right)\right)$$

$$\psi_x(x,y) = X_{12}\cos\alpha x\left(\sin 2\beta y + X_1\left(1 - \frac{2y}{b}\right)\right) \tag{7.29}$$

$$\psi_y(x,y) = Y_{12}\sin\alpha x\cos 2\beta y$$

$$q(x,y) = Q_{12}\sin\alpha x\sin 2\beta y$$

By substituting (7.29) into (7.23), we obtain the stiffness matrix for an ortho-tropic deck under antisymmetric loading (Brown 1998). The constants W_1 and X_1 are determined as

$$W_1 = A_{44}c\left(\frac{Y_{12}}{W_{12}} + 2\beta\right)\left(\frac{1}{1 + \frac{2}{b}A_{44}c}\right) \tag{7.30}$$

$$X_1 = -\frac{A_{44}}{\alpha^3 D}\left(\frac{Y_{12} + 2\beta W_{12}}{X_{12}}\right)$$

where c is the same as given in (7.27). The generalized deflection function for antisymmetric loading is

$$w^R(x,r) = R_{12} \frac{1}{\alpha^2} \left(\frac{1}{\kappa F} + \frac{1}{\alpha^2 D} \right) \sin \frac{\pi x}{a} \left(\sin \frac{2\pi r}{n} + W_1 \cos \frac{\pi r}{n} \right) \tag{7.31}$$

where

$$R_{12} = \frac{Q_{12}}{\frac{1}{\alpha^2} \left(\frac{1}{\kappa F} + \frac{1}{\alpha^2 D} \right) \frac{Q_{12}}{W_{12}} + \frac{4n}{b} \left(1 + \frac{2W_1}{3\pi} \right)}$$

The asymmetric case is obtained by superposition of the symmetric and antisymmetric load conditions. By simply adding the symmetric and antisymmetric responses, the generalized deflection function for an interior stringer under an asymmetric load is written as

$$w^R(x,r) = \left[R_{11} \left(\sin \frac{\pi r}{n} + W_o \right) + R_{12} \left(\sin \frac{2\pi r}{n} + W_1 \cos \frac{\pi r}{n} \right) \right] \frac{1}{\alpha^2} \left(\frac{1}{\kappa F} + \frac{1}{\alpha^2 D} \right) \sin \frac{\pi x}{a} \tag{7.32}$$

7.4.2 Wheel Load Distribution Factors

The above solution is used to define wheel load distribution factors for any of the stringers. The load distribution factor for any interior stringer ith is defined as the ratio of the interaction forces $R(x, r)$ for the ith stringer to the sum of interaction forces for all stringers. The general expressions of load distribution factors in terms of the number of stringers m (where $m = n + 1$) for symmetric and asymmetric loads (Brown 1998) are, respectively,

$$W_f^{Sym}(r) = \frac{\sin \frac{r-1}{m-1}\pi + W_o}{\frac{2}{\pi}(m-1) + mW_o}$$

$$W_f^{Asym}(r) = \frac{\left[R_{11}\left(\sin \pi \frac{r-1}{m-1} + W_o \right) + R_{12}\left(\sin 2\pi \frac{r-1}{m-1} + W_1\left(1 - 2\frac{r-1}{m-1} \right) \right) \right]}{\sum_{r=1}^{m} \left[R_{11}\left(\sin \pi \frac{r-1}{m-1} + W_o \right) + R_{12}\left(\sin 2\pi \frac{r-1}{m-1} + W_1\left(1 - 2\frac{r-1}{m-1} \right) \right) \right]} \tag{7.33}$$

7.4.3 Design Guidelines

Based on the wheel distribution factors obtained above, the number of stringers necessary for a given bridge deck can be determined. The dimensions of the deck are used to evaluate the maximum allowable moment per lane (M_{max}) according to AASHTO (1989). Then, an equivalent concentrated load (P_e) is calculated as (Salim 1997; Brown 1998)

$$P_e = \frac{4M_{max}}{L} \qquad (7.34)$$

where L is the length of a stringer (span of the bridge).

The equivalent deck properties (Section 7.3) and the bending and shear stiffnesses (D and F) for a given type of stringer (Section 7.2) are then used to calculate the edge deflection coefficient W_o or W_1. Next a design load (P_d) is defined for either a symmetric or asymmetric load case as

$$P_d = P_e N_L (W_f)_{max} \qquad (7.35)$$

where N_L is the number of lanes and $(W_f)_{max}$ is found from (7.33) as a function of number of stringers m. Two design criteria based on the performance of stringers and deck can be used to design the system.

7.4.3.1 Design Criterion Based on Performance of Stringer

The mid-span deflection δ_{LL} of a stringer is evaluated as

$$\delta_{LL} = P_d \left(\frac{L^3}{48D} + \frac{L}{4\kappa F} \right) (1 + DLA) \qquad (7.36)$$

where DLA is the dynamic load allowance factor, and for short-span bridges $DLA \cong 0.2$ (Salim et al. 1997). Equation (7.36) is then set equal to the maximum allowable deflection (from AASHTO 1989) to determine the number of stringers required for the bridge deck. Once a suitable system is chosen, the maximum moment due to live load (M_{LL}) is calculated from

$$M_{LL} = \frac{P_d L}{4} (1 + DLA) \qquad (7.37)$$

Finally, the approximate maximum extreme fiber normal stress (σ_c) in the stringer can be found from

$$\sigma_c = \frac{M_{LL} y'}{I} \qquad (7.38)$$

where y' is the distance from the neutral axis of the stringer to the top surface of the stringer and I is the moment of inertia of the stringer. This stress can then be compared with the material compressive or tensile strength to confirm that the system will be effective. Also, as an approximation, shear stress in the stringer can be estimated as

$$\tau = \frac{P_d}{2A_w}(1+DLA) \tag{7.39}$$

where A_w is the area of the web panels. The shear stress in (7.39) should be less than the shear strength of the stringer.

7.4.3.2 Design Criterion Based on Performance of Deck

Excessive local deck deformation and punching-shear failure may be observed in FRP bridge applications. Thus it is necessary in the design process to check the local deck deflection and bending and shear stresses in a deck section between two adjacent stringers (Davalos and Salim 1993, 1995). Further research is needed to address these issues.

7.4.4 Experimental Testing and Numerical Analysis of FRP Deck-Stringer Systems

To validate the approximate series solution presented above, an FRP deck 3.048 m × 3.048 m × 20.32 cm (10 ft × 10 ft × 8 in.) is fabricated by bonding side-by-side box beams of 10.16 × 20.32 × 0.635 cm (4 × 8 × ¼ in.) (Figure 7.2); the deck is attached to FRP I-beams 30.48 × 30.58 × 1.27 cm (12 × 12 × ½ in.) (Figure 7.3) and tested and analyzed as a deck-stringer bridge system. The

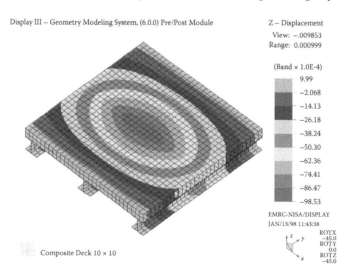

FIGURE 7.15
FE simulation and deflection contour of deck-stringer system under symmetric load.

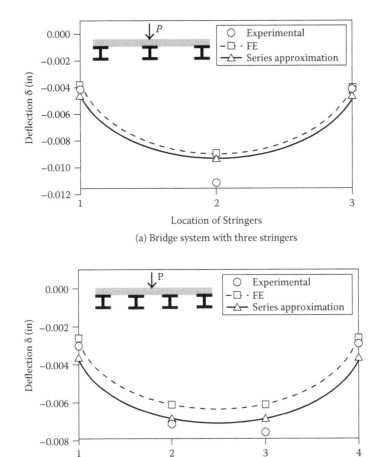

(a) Bridge system with three stringers

(b) Bridge system with four stringers

FIGURE 7.16
Comparison for deck-stringer bridge system.

deck with either three or four stringers is subjected to various static load conditions (Brown 1998). The finite element model with NISA (1994) is shown in Figure 7.15 for a three-stringer system under a concentrated centric loading. The comparisons among the FE, series solution, and experiments for both three-stringer and four-stringer systems under centric loading are shown in Figure 7.16, and relatively consistent trends are observed. The maximum differences of stringer deflections between experiments and series approximation are about 16% for a three-stringerr system and 23% for a four-stringer system. The detailed study on the experimental program and comparisons for various cases can be found in Brown (1998).

7.5 Design Analysis Procedures and Illustrative Example

General guidelines for the applications of the above series approximation solution for design analysis of FRP composite deck-and-stringer bridge systems and several illustrative design examples are given in Brown (1998).

7.5.1 General Design Procedures

The following step-by-step design procedures are recommended:

1. Define bridge dimensions and allowable loads.
2. Obtain deck panel and bridge stringer properties by micro/macromechanics (Luciano and Barbero 1994; Davalos et al. 1996a; Barbero 1999) and mechanics of laminated beams (MLB) (Barbero et al. 1993).
3. Determine deck equivalent material properties by the equations derived in Section 7.3.
4. Perform the series approximation analysis and determine the number of stringers m based on the required deflection limit.
5. Check the stress level on the stringers.
6. Check the local stresses, deflections, and other details.

7.5.2 Design Example

As an illustration, a single-lane short-span bridge of 0.48 m (15 ft) width and 0.79 m (25 ft) span is designed using side-by-side $10.16 \times 20.32 \times 0.635$ cm ($4 \times 8 \times \frac{1}{4}$ in.) bonded FRP box sections for the cellular deck assembly and optimized FRP winged box beams 30.48×60.96 cm (12×24 in.) (Qiao et al. 1998; Davalos et al. 1996b) for the stringers (Figure 7.17). The material properties for the stringers are $D = 3.590 \times 10^7$ N-m^4/m^2 (1.248×10^{10} lb-in^4/in.^{2}) and $F = 8.651 \times 10^7$ N-m^2/m^2 (1.940×10^7 lb-in.^{2}/in.^{2}), which are computed by micro/macromechanics and MLB. The deflection limit of L/500 and the loading of AASHTO HS20 (1989) are considered, and the number of stringers (m) is used as a design variable. The edge deflection coefficient for symmetric loading is evaluated from (7.27) as $W_o = 1.691$, and the deflection limit is written as a function of the number of stringers m:

$$\delta_{LL} = \frac{L}{500} = M_{max}\left(\frac{4}{L}\right)N_L\left(\frac{1+W_0}{mW_0+(m-1)\frac{2}{\pi}}\right)\left(\frac{L^3 48}{D}+\frac{L}{4\kappa F}\right)(1+DLA) \qquad (7.40)$$

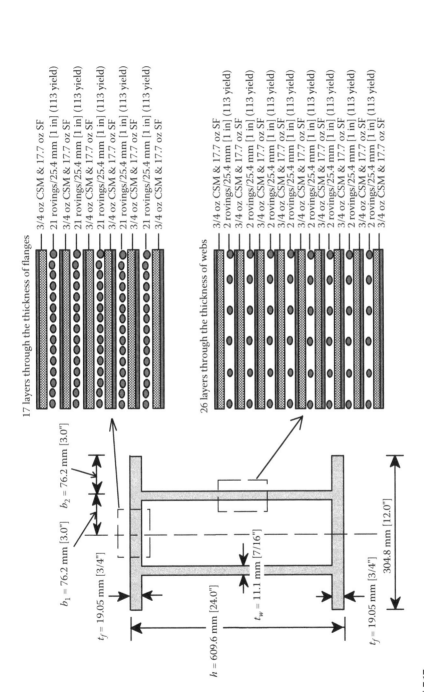

FIGURE 7.17
Dimensions and panel fiber architectures of optimized winged box beam. (From Qiao, P., Davalos, J. F., and Barbero, E. J., *Journal of Composite Materials*, 32(2), 177–196, 1998.)

where $N_L = 1.0$ and $\kappa = 1.0$. In this example, the dynamic load allowance $DLA = 0.20$ and AASHTO lane moment $M_{max} = 28{,}188$ kN-m (207.4 kip-ft) are used. Solving for the number of winged box beams (stringers) (m) required for this single-span bridge, we get $m = 6.15$, and therefore $m = 7$ is used, which corresponds to 76.2 cm (30 in.) center-to-center spacing of seven longitudinal stringers. The maximum stress in the stringer becomes $\sigma_c = 1.02$ kN/cm² (1.48 ksi), which is below the allowable stress of 14.64 kN/cm² (21.2 ksi) (Qiao et al. 1998).

7.6 Conclusions

As described in this chapter, a systematic approach for design analysis of FRP deck-stringer bridge systems is proposed, and the constitutive material properties and micro/macrostructure of a composite system are accounted for in the design. This design approach (Figure 7.1) includes the analyses of ply (micromechanics), panel (macromechanics), beam or stringer (mechanics of laminated beam), deck (elastic equivalence model), and finally, combined deck-stringer system (series approximation technique). This relatively simple and systematic concept accounts for the complexity of composite materials and the geometry of the bridge system. The approximate series solution, which is used to obtain wheel load distribution factors for symmetric and asymmetric loading, is an efficient way to analyze and design single-span FRP deck-stringer systems. The present design analysis approach can be efficiently used to design bridge systems and also develop new design concepts for single-span FRP deck-stringer bridges.

References

AASHTO. (1989). *Standard specifications for highway bridges*. American Association of State Highway and Transportation Officials, Washington, DC.

Barbero, E.J. (1999). *Introduction to composite materials design*. Taylor & Francis, Philadelphia, PA.

Barbero, E.J., Lopez-Anido, R., and Davalos, J.F. (1993). On the mechanics of thin-walled laminated composite beams. *Journal of Composite Materials*, 27(8), 806–829.

Brown, B. (1998). Experimental and analytical study of FRP deck-and-stringer short-span bridges. Master of science thesis, West Virginia University, Morgantown.

Cusen, A.R., and Pama, R.P. (1975). *Bridge deck analysis*. John Wiley & Sons, New York.

Davalos, J.F., and Salim, H.A. (1993). Effective flange-width of stress-laminated T-system timber bridges. *ASCE Journal of Structural Engineering*, 119(3), 938–953.

Davalos, J.F., and Salim, H.A. (1995). Local deck effects in stress-lamianted T-system timber bridges. *International Journal of Structural Engineering Review*, 5(1), 1143–1153.

Davalos, J.F., Salim, H.A., Qiao, P., Lopez-Anido, R., and Barbero, E.J. (1996a). Analysis and design of pultruded FRP shapes under bending. *Composites: Part B: Engineering Journal*, 27(3–4), 295–305.

Davalos, J.F., Qiao, P., and Barbero, E.J. (1996b). Multiobjective material architecture optimization of pultruded FRP I-beams. *Composite Structures*, 35, 271–281.

Davalos, J.F., Qiao, P., Barbero, E.J., Troutman, D., and Galagedera, L. (1998). Design of FRP beams in engineering practice. In *Proceedings of International Composites Expo '98*, Composites Institute, pp. 12-E (1–6).

Luciano, R., and Barbero, E.J. (1994). Formulas for the stiffness of composites with periodic microstructure. *International Journal of Solids and Structures*, 31(21), 2933–2944.

Numerically Integrated Elements for System Analysis (NISA). (1994). *Users manual.* Version 94.0. Engineering Mechanics Research Corp., Troy, MI.

Qiao, P., Davalos, J.F., and Barbero, E.J. (1998). Design optimization of fiber-reinforced plastic composite shapes. *Journal of Composite Materials*, 32(2), 177–196.

Qiao, P.Z., Davalos, J.F., and Brown, B. (2000). A systematic approach for analysis and design of single-span FRP deck/stringer bridges. *Composites: Part B: Engineering Journal*, 31(6–7), 593–609.

Reddy, J.N. (1984). *Energy and variational methods in applied mechanics.* John Wiley, New York.

Salim, H.A. (1997). Modeling and application of thin-walled composite beams in bending and torsion. PhD dissertation, West Virginia University, Morgantown.

Salim, H.A., Davalos, J.F., Qiao, P., and Barbero, E.J. (1995a). Experimental and analytical evaluation of laminated composite box beams. In *Proceedings of 40th International SAMPE Symposium*, vol. 40, no., pp. 532–539.

Salim, H.A., Davalos, J.F., GangaRao, H.V.S. and Raju, P. (1995b). An approximate series solution for design of deck-and-stringer bridges. *International Journal of Engineering Analysis*, 2, 15–31.

Salim, H.A., Davalos, J.F., Qiao, P., and Kiger, S.A. (1997). Analysis and design of fiber reinforced plastic composite deck-and-stringer bridges. *Composite Structures*, 38, 295–307.

Troitsky, M.S. (1987). *Orthotropic bridges, theory and design.* James F. Lincoln ARC Welding Foundation, Cleveland, OH.

Index